Analytic Idealism in a Nutshell

A straightforward summary of the 21st century's only plausible metaphysics

Analytic Idealism in a Nutshell

A straightforward summary of the 21st century's only plausible metaphysics

Bernardo Kastrup

IFF
BOOKS

London, UK
Washington, DC, USA

CollectiveInk

First published by iff Books, 2024
iff Books is an imprint of Collective Ink Ltd.,
Unit 11, Shepperton House, 89 Shepperton Road, London, N1 3DF
office@collectiveinkbooks.com
www.collectiveinkbooks.com
www.iff-books.com

For distributor details and how to order please visit the 'Ordering' section on our website.

Text copyright: Bernardo Kastrup 2023

ISBN: 978 1 80341 669 4
978 1 80341 671 7 (ebook)
Library of Congress Control Number: 2023945574

A CIP catalogue record for this book is available from the British Library.

Design: Lapiz Digital Services

UK: Printed and bound by CPI Group (UK) Ltd, Croydon, CR0 4YY
Printed in North America by CPI GPS partners

We operate a distinctive and ethical publishing philosophy in all areas of our business, from our global network of authors to production and worldwide distribution.

Previous Books

Rationalist Spirituality: An exploration of the meaning of life and existence informed by logic and science
Iff Books, ISBN: 978 1 84694 407 9

Dreamed up Reality: Diving into mind to uncover the astonishing hidden tale of nature
Iff Books, ISBN: 978 1 84694 525 0

Meaning in Absurdity: What bizarre phenomena can tell us about the nature of reality
Iff Books, ISBN: 978 1 84694 859 6

Why Materialism Is Baloney: How true skeptics know there is no death and fathom answers to life, the universe, and everything
Iff Books, ISBN: 978 1 78279 362 5

Brief Peeks Beyond: Critical essays on metaphysics, neuroscience, free will, skepticism and culture
Iff Books, ISBN: 978 1 78535 018 4

More Than Allegory: On religious myth, truth and belief
Iff Books, ISBN: 978 1 78535 287 4

The Idea of the World: A multi-disciplinary argument for the mental nature of reality
Iff Books, ISBN: 978 1 78535 739 8

Decoding Schopenhauer's Metaphysics: The key to understanding how it solves the hard problem of consciousness and the paradoxes of quantum mechanics
Iff Books, ISBN: 978 1 78904 426 3

Decoding Jung's Metaphysics: The archetypal semantics of an experiential universe
Iff Books, ISBN: 978 1 78904 565 9

Science Ideated: The fall of matter and the contours of the next mainstream scientific worldview
Iff Books, ISBN: 978 1 78904 668 7

Contents

Chapter 1

What is this book about?

This is a book about the nature of reality. It elaborates on the best hypothesis we have today, based on leading-edge science and analytic reasoning, about what reality *is*. Is the world physical, in the sense of consisting of mass, charge, frequency, and other physical quantities? If not, is it instead informational, in the sense of being constituted by purely abstract mathematical patterns? Or is it perhaps mental, in the sense of being made of experiential qualities? These are the questions this book attempts to tackle.

Notice that such questions of *being*—as in what reality *is*—cannot be definitively answered by science alone; those who think they can have a fundamentally flawed understanding of both science and philosophy. The scientific method can only definitively answer questions of *behavior*: what nature *does*, as opposed to what it *is*. After all, science is based on controlled empirical experimentation: it poses a question to nature in the form of an experiment, which nature then answers by *doing* something in response. This *doing* is a behavior of nature, not an unambiguous expression of what nature is, as different hypotheses about the essence of reality may be equally consistent with the observed behavior. Moreover, the litmus test of scientific theories is their ability to predict what nature will *do next*, which again is a question of behavior.

As such, although the argument in this book is certainly informed by science—in that some empirically observed natural behaviors are inconsistent with, and hence allow us to rule out, certain hypotheses about the nature of reality—it is not a purely scientific one. Indeed, in addition to science, this book also leverages the methods of philosophy, particularly metaphysics,

1

the area of philosophy dedicated to questions of being. More specifically, beyond the empirical adequacy of the hypothesis it puts forward, this book also leverages softer truth guidelines such as conceptual parsimony, internal logical consistency, overall coherence, and explanatory power. These guidelines cannot lead us to definitive answers to questions of being, but surely allow us to rank the hypotheses at hand and figure out the most likely one. This is the spirit of this book.

Analytic Idealism — the metaphysical hypothesis that gives this book its title, and which is elaborated upon in the next chapters — acknowledges that there is an external world out there, beyond our individual minds. It also acknowledges that this external world unfolds spontaneously, according to its own inherent dispositions, which in turn lead to regularities of behavior that we've come to call the 'laws of nature.' Analytic Idealism then acknowledges that human reason can recognize and model these regularities, thereby being able to predict nature's future behavior. Finally, it acknowledges that complex natural phenomena can be sufficiently accounted for in terms of simpler ones.

This way, Analytic Idealism embraces *realism* (i.e., there is an external world out there, independent of our individual minds; independent of our observation, volition, fantasies, preferences, rituals, etc.), *naturalism* (i.e., the phenomena of the external world unfold spontaneously, according to nature's own inherent dispositions, and not according to external intervention by a divinity outside nature), *rationalism* (human reason can recognize and model the regularities of nature's behavior), and *reductionism* (complex phenomena can be explained in terms of simpler ones).

However, Analytic Idealism then infers that the external world is of the same ontic *kind* or *essence* as our individual minds. In other words, it posits that the world out there is *mental* (in the sense of being experiential, but *not* in that it comprises high-level, human-like cognition), identical in essence to our own individual thoughts and emotions, although not constituted by our thoughts and emotions. This is analogous to how my thoughts are mental just as yours are too, yet mine are external to *you* and not part of *your* mental inner life. My thoughts would still be here, doing whatever it is that they do, even if you were not around, and regardless of your particular preferences. From *your* perspective, my thoughts are external and objective; even though, from their own perspective, they are subjective, and thus mental. Now, in exactly the same way, Analytic Idealism posits that the world out there is thought-like in essence, but nonetheless external to *our* individual minds. The world is subjective from its own perspective, but objective from *ours*. The world does what it does, according to its own spontaneous and inherent natural dispositions, regardless of whether we like it or not, or even of whether we are here to observe it or not. And yet it is, in itself, mental, subjective, experiential.

Clearly, Analytic Idealism is very different from the mainstream metaphysical hypothesis of the early 21st-century: Physicalism. The latter posits that the essence of the world out there is not mental at all but, instead, physical in a strict sense. What does it mean to say that reality is strictly physical in essence? It means that reality should be, in principle, exhaustively characterizable through *quantities* and their mathematical relationships alone. In other words, according to mainstream Physicalism, if you come up with a long enough list of the right numbers—such as the amount of mass and charge at different coordinates of spacetime, or the amplitude and frequency of different field oscillations, etc.—and associated equations, you will have said *everything* there is to say about reality; nothing will remain uncharacterized or ambiguous.

This is clearly antithetical to Analytic Idealism, according to which nature is made of *felt qualities*—not only the qualities you, I, and other living beings happen to feel, but felt qualities nonetheless—as opposed to abstract quantities. In other words, under Analytic Idealism there is something it is like to *be* the world out there; the world is made of transpersonal experiential states that cannot be exhaustively characterized in terms of quantities alone. After all, how do we fully characterize, using numbers alone, what it feels like to, e.g., fall in love? What *quantities* can exhaustively capture the *quality* of falling in love? We cannot know what it feels like to fall in love merely by reading an extensive list of numbers and equations; we can only know it by direct, *qualitative* acquaintance. Clearly, thus, Analytic Idealism entails a fundamentally different understanding of the nature of reality than mainstream Physicalism.

I will elaborate on all this much more extensively in future chapters, so don't be discouraged if you can't wrap your head around it right now. At this point, I'm only trying to give you a general sense of where I am going with all this, to preempt the possibility that an early misinterpretation or prejudice might cloud your ability to grasp my intended meaning.

The discussion above raises an immediate question: how plausible is it that our culture, having put a man on the moon, cured countless diseases, changed our very genes, and created the Internet, is so fundamentally wrong about the essence of the world we manipulate so effectively through our technology? How plausible is it that, having accomplished so much, we are still so mistaken about the basics?

It is more than plausible; it is certain. And the ones among us looking carefully and thoughtfully at the state of play in science and philosophy know this.

I shall discuss this untrivial claim at length throughout this book. For now, though, notice that technology necessitates no ultimately correct understanding of the nature of reality; it only needs *empirically convenient fictions*. To see how, consider that a 5-year-old kid can be world champion playing a computer game without having the slightest idea of what the game actually *is* — that is, of the computer hardware and software that constitute the game. To be world champion, all the kid needs is a convenient fiction in terms of which to relate to the game. And it may go like this: there is a little man inside the screen; I am that little man; if I shoot those other little men in the screen, I score points; if I get shot or touch this or that wall, I die; and so on. Each and every element of this fiction is utterly false: there is no little man inside the screen; you are not inside the screen; you are not shooting, or getting shot by, anyone; there are no walls to touch; and you don't die from playing the game. Yet, the fiction is convenient in that the game behaves, for all applicable purposes, *as if* the fiction were true; and that's all the kid needs to play it well and be crowned world champion.

Technological prowess is entirely analogous to this: all it needs are empirically convenient fictions in terms of which to relate to reality. For instance, we have put a man on the moon under the convenient fiction that there are invisible forces, acting instantaneously and at a distance, attracting a spacecraft to both the Earth and the Moon, in a kind of celestial tug of war. We even gave these forces a name: gravity. Needless to say, since Albert Einstein's general theory of relativity we have known that there are no such forces; what we call gravity is but a curvature of the invisible fabric of spacetime — our new empirically convenient fiction — not a force. And although the designers of the Apollo missions knew this, the earlier convenient fiction — Newtonian mechanics — was a good enough empirical approximation for their purposes, and so they used it in their calculations. That fiction was convenient enough for the purposes of their game, and that's all that is required for technological advancement.

Engineers know that, not only is it unnecessary for our convenient fictions to be true — finite-element modeling is not true, Fourier optics is not true, discrete-element transmission line theory is not true, even the notion of subatomic particles is not true, yet we apply them every day because they work well enough in empirical practice—often we don't even need to have a convenient fiction at all. In electronics and telecommunications engineering, for instance, what we refer to as the 'signal-to-noise ratio' is an attempt to quantify the significance of all the unknown effects for which we don't have an explicit convenient fiction (the 'noise'), in relation to those for which we do (the 'signal'). As long as the signal-to-noise ratio is above a certain threshold, our technology will work, for it is engineered precisely to keep the unknowns—which are legion—at bay. As such, that our technology works doesn't mean even that we have a fiction for everything of relevance; let alone that the fictions we do have are true. Convenient fictions really are just that: convenient; their relationship to truth is tenuous.

It is therefore naïve to think that technological success reflects metaphysical understanding, just as it is naïve to think that the 5-year-old world champion actually understands what is going on while he plays the computer game. Add to this the historical fact that every generation before us has been demonstrably wrong about a great many things of relevance regarding the nature of reality, and you will realize how plausible it is that we, too, are wrong when it comes to mainstream Physicalism. Future generations will know this with crystal clarity, just as we today know of the mistakes of our ancestors.

Indeed, it would be silly to think that mere human beings, bipedal primates who have been running around this space rock called Earth for only about two or three hundred thousand years, and sporting an intellect—i.e., the capacity to think symbolically, conceptually—for less than fifty thousand, have evolved a cognitive system advanced enough to comprehend every salient aspect of reality. Of course we haven't; for the same reason that ants haven't evolved enough to comprehend Quantum Field Theory. As such, we cannot hope to unveil the ultimate truths, and cannot be certain of any of our fictions.

But there is something we *can* be certain of: *our current and past mistakes*. For these mistakes entail internal logical contradictions, demonstrable empirical inadequacies, or fundamental gaps in explanatory power that rule them out as viable hypotheses. We very well *can* discern these errors with certainty, for they are *our* errors—the expression of our own primate foolishness—not genuine natural mysteries. So the game is to unveil our mistakes, correct them through new convenient fictions, and thereby get clo*ser* to a viable understanding of the nature of reality. The hope is not that monkeys can discern the ultimate truths of existence, but that monkeys can be *less* wrong over time. This we *can* do; and this we *must* do, for acquiescing to known errors is morally unacceptable.

To take steps forward, however, we must face up to the fact that mainstream Physicalism is not only a fiction, but one that isn't even convenient anymore. In the early days of the Enlightenment, it did serve a sociopolitical purpose as the tensions between a nascent science and the Church grew. By carving out a metaphysical domain outside 'spirit'—a translation of the Greek word 'psyche,' which also means 'mind'—early scientists hoped to be able to operate without being burned at the stake, as Giordano Bruno was in 1600. The notion of a physical world fundamentally different from, and outside, psyche must have sounded ludicrous and harmless enough to Church authorities at the time that they left scientists alone.

As a matter of fact, it isn't a secret that the founders of the Enlightenment were well aware that Physicalism was a political weapon first and foremost, not a plausible account of the nature of reality. Will Durant, in *The Story of Philosophy* (Simon and Schuster, 1991), points out that Denis Diderot—one of the authors of the *Encyclopédie*, the founding document of the Enlightenment—acknowledged that "all matter is probably instinct with life, and it is impossible to reduce the unity of consciousness to matter and motion; but materialism is a good weapon against the Church, and must be used till a better one is found" (page 300). Diderot had clarity and honesty about what he was up to, which we can't say of most self-appointed spokespeople of Physicalism and 'Scientism'—a naïve and fallacious conflation of science and metaphysics—today.

Later, in the second half of the 18th century, the industrial revolution—with its railways, steam engines and machine tools—was picking up steam and the emergent commercial class, the bourgeoisie, was accruing social influence. As a consequence of this process, by the time of the July Revolution of 1830 the zeitgeist of the Enlightenment had shifted from edifying art and philosophy to down-to-earth technology and the practical applications of science. At this point, the same notion of a physical realm distinct from psyche not only justified the growing dominance of bourgeois intellectual elites over the clergy, but also provided a psychological handle—a convenient fiction—to help scientists extricate themselves from the phenomena they observed. This, in turn, may have helped increase the objectivity of empirical experimentation at a crucial early juncture for science.

But by the second half of the 19th century, the original Enlightenment clarity that Physicalism was mostly a political weapon—instead of a truly plausible metaphysical hypothesis— had been lost, as chronicled by Charles Taylor in *A Secular Age* (Harvard University Press, 2007). Yet this, too, came with a payoff; in fact, the biggest psychological payoff of all: understanding phenomenal consciousness—i.e., our very ability to experience—as a mere by-product of physical arrangements eliminated, in one fell swoop, *the single greatest fear humankind had had throughout its history*; namely, the fear of what we will experience after death. For if our minds are generated by living brains, then there will be no consciousness to experience anything after death. The angst of the great unknown—codified in Christian mythology as the fear of Hell—was suddenly gone, along with all the moral responsibility that had hitherto oppressed the lives of Christians. All of our worries, regrets, anxieties, etc., were now *guaranteed*—whether we liked it or not, believed it or not—to come to an end at the moment of death. It is hard for us today to imagine what a profound liberation this must have felt like, for we now take this liberation entirely for granted. Perhaps only the Reformation, a couple of centuries prior, could be compared to it. The fear that had allowed the Church to singlehandedly control practically the entire population of the European continent for over a millennium was suddenly off the table. It's no wonder that Physicalism accrued so much cultural momentum ever since, despite being perhaps the worst metaphysical hypothesis—in terms of explanatory power, conceptual parsimony, empirical adequacy, internal consistency, and overall coherence—ever to gain mainstream status in any society on the planet.

The historical, sociopolitical, and psychological convenience of the physicalist fiction is almost impossible to overestimate. Through most of the 20th century, it was even used to justify *itself*, in a surreal feat of circular reasoning. For the greatest embarrassment of mainstream Physicalism is its fundamental inability to account for phenomenal consciousness—i.e., the existence of experience—which is nature's sole pre-theoretical, given fact. Indeed, phenomenal consciousness precedes theory epistemically, in that all theories arise and exist within it. But by pronouncing phenomenal consciousness an epiphenomenon of the physical world—a mere by-product of brain processes, devoid of causal powers of its own—without any explanation as for how this could possibly be so, mainstream Physicalism eliminated consciousness from the agenda of human investigation; it suppressed the very thing that could expose its own most fundamental shortcoming. To put it simply, the idea was that, since we assume that consciousness is epiphenomenal and absent from the causal nexus anyway, we don't really need to bother about finding out how it comes about—how utterly convenient! One can clearly see this circularity in the development of, e.g., Positivism and Behaviorism in the 20th century. And it took until 1974, with Thomas Nagel's seminal paper titled "What is it like to be a bat?", for consciousness to (very) slowly return to the investigative agenda.

Our culture's confidence in mainstream Physicalism is a historical psychosocial phenomenon largely unrelated to reason and evidence. The only reason most of us don't see this is that our lives are too short for us to discern the relatively slow, subtle ebb and flow of cultural history in which we are immersed. We spend almost the entirety of our ephemeral existences surrounded by learned elites who seem very confident in their internally inconsistent, empirically untenable, and explanatorily hopeless views. And thus, we think that mainstream Physicalism *must* be true; how could all these people be wrong? Yet, history shows that 'all these people' have *always* been wrong before, while reason and evidence show that they are wrong now too. Already in the next chapter, we shall see precisely how.

Analytic Idealism—the subject of this book—represents a correction of our known metaphysical mistakes; a step forward. As I shall soon argue, it offers the most plausible and parsimonious hypothesis we have today about the nature of reality. Herein lies the value of what you are about to read.

I have written ten earlier books and a PhD thesis on the subject, not to mention a number of technical papers in academic journals, blog posts, and popular science & philosophy essays in major publications. So, what is new in this particular volume? As the title of this book indicates, here I attempt to summarize, in an informal but direct manner, the key salient points of Analytic Idealism and the argument that substantiates it. Ideas from several of my previous writings are revisited here, but often in a new form, from a different slant. And they are brought together so to give you the briefest and most compelling overview of Analytic Idealism I could muster.

In addition, as I've found myself having to explain and defend Analytic Idealism in countless interviews, Q&A sessions, panels, debates, courses, and other public events over the years, I've had to distill a more optimal way to bring forth the core ideas. I've learned over time what the main difficulties are that different people have with Analytic Idealism, and refined ways to explain it so to meet people where they are, honoring their intuitions and tackling their hidden assumptions more explicitly. All these learnings and refinements are built into the present volume.

Stylistically, my previous ten books were meticulously documented. The same goes for my second PhD thesis and my many technical papers in peer-reviewed academic journals. I thus believe that I have earned the right to discuss Analytic Idealism now in a less formal, less documented, but more fluent and easy-to-read manner, capturing the most salient points in more intuitive, colloquial language. This is what I try to achieve in this book. Unlike previous writings, here I shall thus deliberately avoid formal literature citations, bibliography, and notes. Whenever a literature reference seems particularly productive or unavoidable, I shall mention it in the running text, just as I already did above.

This book is meant to be as close as possible to a verbal discussion of Analytic Idealism, as if I were explaining it to you in person. The tone adopted deliberately reveals more aspects of my own humanity and emotional state to the reader, which can be contrasted to the drier and more objective character of my technical writings. For those readers who prefer or require a technical and more rigorously documented argument, I recommend my earlier output, much of which—such as the academic papers and thesis—is freely available online.

Chapter 2

What you see is *not* what you get

When we look around, we see a world of objects with particular shapes, as well as events unfolding in defined locations. We then automatically assume that the world around us is constituted by these objects and events; by these shapes and locations. In other words, we believe that our perception is a kind of *transparent window into the world*, revealing to us the world *as it is in itself*.

But can we conceive of perception as something other than a transparent window into the world? Of course we can. Take an airplane, for instance: it has a number of sensors—such as air speed, pressure, and orientation sensors—which measure the states of the sky outside the airplane. The resulting measurements are then displayed to the pilot in an encoded manner, in the form of dial indications in the airplane's dashboard. As such, the airplane's dashboard conveys accurate and important information about the sky outside, albeit in an encoded form. The pilot must take this encoded information seriously, lest the airplane crashes.

Nonetheless, the dashboard looks nothing like the sky outside: dozens of dials on a flat instrument panel look very different from the 3D clouds, air flow patterns, precipitation, and distant horizon outside the airplane. Clearly, that the dashboard conveys accurate and relevant information about the sky doesn't entail or imply that it must *look like* the sky; as a matter of fact, it doesn't, and we all intuitively understand why this isn't a problem.

Just like the airplane, we, too, are equipped with sensors to collect information about the world surrounding us: our retinas, eardrums, taste buds, mucous lining of the nose, and outer surface of the skin make measurements of the states of the environment surrounding us. The results of these measurements are then represented on the screen of perception: what we see, hear, touch, taste, and smell. As such, the screen of perception is entirely analogous to the airplane's dashboard: both display information about our environment that was collected by our sensors.

Yet, as we've just seen, the airplane's dashboard does not need to look like the sky to convey accurate, relevant, and actionable information about the sky. By the same token, neither does our screen of perception: its contents do *not* need to look like the world in order to convey accurate, relevant, and actionable information about the world. Put in a different way, to do its job properly the screen of perception does not need to be a transparent window into the world; instead, it can be just like the dashboard of dials in an airplane, which looks nothing like the sky outside. The world around us, as it actually is in itself, independently of perception, can in principle be as different from what we perceive as the clouds in the sky are different from dial indications on an airplane's dashboard. What we see is not necessarily what we get; it doesn't need to be.

Let us now ask a more consequential question: although the world doesn't *need* to look like the contents of perception, *can* it look like what we perceive? *Can* perception be a transparent window into the world, even if it doesn't need to?

The answer is a categorical 'No': perception *cannot* be a transparent window into the world. For if it were, our internal cognitive states would need to *mirror* the external states of the world, as this is what is entailed by the 'transparent window' hypothesis. But since there is no *a priori* upper bound to the dispersion of states of the world — i.e., the entropy of the world — there would also be no upper bound to the dispersion of our internal cognitive states, insofar as the latter mirror the states of the world. In other words, if perception were a transparent window allowing us to see the world as it actually is, there would be no upper bound to our internal entropy. Physically, this means that by merely looking at the world we could literally melt into hot meat soup; for unbound internal entropy is incompatible with structural and dynamical integrity. And since we have never witnessed anyone spontaneously melt after taking a look around, it is safe to say that perception does *not* mirror the states of the world; it is *not* a transparent window that allows us to see the world as it is. Why would it?

Instead, perception is like a dashboard of dials: it contains information about the world only in an *encoded* form, thereby limiting the entropy of our internal states, just as a dashboard of dials intrinsically limits the entropy of information pilots have to contend with (incidentally, this is why flight manuals, despite being thick, are finite in length and do not have to tell the pilot what to do for every possible configuration of the sky). All this has been worked out rigorously by Karl Friston and his collaborators, particularly in a paper called "Cognitive dynamics: From attractors to active inference," published in the *Proceedings of the IEEE* in 2014.

Another team has arrived at the same conclusion independently, through an entirely different line of reasoning based not on thermodynamics, but game theory instead. Our perceptual apparatus has evolved for survival fitness, not to show us the world as it actually is; the latter is entirely irrelevant as far as natural selection is concerned. In other words, we evolved to perceive the world in whatever encoded manner helps us react most effectively to environmental challenges and survive. Perceiving the world as it actually is isn't effective for survival, as Prof. Donald Hoffman and his collaborators, from the University of California at Irvine, have been showing mathematically for many years now.

To gain intuition into why a transparent window into the world would "swiftly lead to our extinction," as Prof. Hoffman likes to put it, consider one of his favorite analogies: that of a computer desktop. In it, your files are represented as icons: little colorful rectangles. But do your files, as they are in themselves, actually look like colorful rectangles? Not at all. Computer files are giant sets of millions of microscopic electronic switches, some open and some closed, in non-volatile memory chips made of silicon, metals and oxides. Little could look more different from these files than rectangular icons on a screen. But if we were forced to see our computer files as they actually are— i.e., as millions of microscopic electronic switches—it would become impossible for us to use the computer; we wouldn't be able to make heads or tails of what we see. Instead, we would be overwhelmed by non-actionable information, in which the salient properties of the files would be indiscernible.

To avoid this dysfunctional scenario, the computer's operating system represents the files *not* as they actually are, but in an *encoded* form that conveys what is salient about them in an actionable manner; i.e., as little rectangular icons. The same goes for perception: if evolution had built our perceptual system to cognize the world as it really is, it stands to reason that we would be driven to extinction because of non-actionable information overload. Instead, our perceptual system captures and *represents* the world in an encoded and actionable manner. The contents of perception—the objects, shapes, and events we see around us—are mere 'icons' in the 'graphical user-interface' entailed by the screen of perception. We never become directly acquainted with the world as it actually is, because doing so would make it more difficult for us to compete and survive in our ecosystem.

We are airplane pilots who were born in the cockpit of an airplane without windows; an airplane that can only be flown by instruments. We have never left the cockpit to see the world outside as it actually is. All we have ever been acquainted with is the dashboard of dials that we call 'perception.' It is thus psychologically understandable that we should mistakenly take the dashboard for the world: we've never had anything else. Even our language and our very thoughts have evolved to 'speak dashboard,' not 'reality.' In other words, because our interactions with the world are *always* mediated by the dashboard, we think and speak in terms of the parameters and scales of the dashboard, not in terms of the actual elements of reality. The entire paradigm of our worldview tends to be the paradigm *of the dashboard*; it follows the way information is organized and displayed by the dials, not the ontic structure of the world itself.

To clearly grasp this is the first necessary step in the journey towards a less-wrong understanding of reality. Every time you look around and see a world of 'physical' — in the colloquial, not the strict sense — objects and events, *you are not seeing reality as it actually is*; all you are seeing is a dashboard representation of reality, constructed by evolution for the benefit of your survival. Reality, as it is in itself, stands to be as different from what you perceive as the clouds in the sky are different from the dial indications on an airplane's dashboard.

Taking the contents of perception for reality is as silly a mistake as a pilot who looks at his dashboard and declares it to *be* the sky outside. Just as the dashboard conveys accurate and important information about the sky outside without *being* the sky, so does perception convey accurate and important information about the reality outside without *being* that reality. You *never* get to see reality; all you see is an encoded representation thereof, meant to convey actionable information about your environment, while limiting your internal entropy and improving your chances of reacting effectively to environmental challenges and opportunities.

What we colloquially refer to as the 'physical' world—i.e., the perceived world of objects, colors, shapes, and events around us, occupying the scaffolding of spacetime—*is the dashboard*, not reality as it is in itself. 'Physicality' in the colloquial sense—i.e., the contents of perception—is an *internal cognitive representation* of reality, which we create ourselves, given the cognitive apparatus evolution has equipped us with. There surely is a *real* world out there, independent of perception; a world that would still be there even if we were not here to perceive it; a world that doesn't care whether we like it or not, believe it or not, wish it to be different or not; a world that doesn't change merely because we fantasize about its being different. But that *real* world is not the 'physical' world that appears on the screen of perception. The latter is a mere *representation* of the real world, which *we* create. The thing that is represented is not created by us, but the representations themselves are.

As such, does the moon exist when nobody is looking at it? If by 'the moon' we mean the aspect of reality that is *represented as* that shiny white disk up in the night sky, then yes, it exists whether or not anyone or anything is measuring or observing it. But if by 'the moon' we mean that shiny white disk itself—the colloquially 'physical' entity up in the sky—then no, it doesn't exist if no living being is observing it, for the experiential qualities 'shiny,' 'white,' and 'disk-shaped' are cognitive representations created by the act of observation.

An analogy may help make the point clearer: when there are no airplanes up in the sky, there are no dashboard representations; nothing is being measured and displayed on a dashboard. Similarly, when there are no living beings observing reality, there is no 'physical' world, for the 'physical' world is a set of perceptual representations arising from observation. But none of this means that there is no sky in the absence of airplanes; the clouds are still there; the states of the sky—its air pressure, wind direction, etc.—are all still there, even when not being measured and displayed on a dashboard. Analogously, the reality that *would* be represented as contents of perception — i.e., as 'physical' objects and events — in case someone were observing it exists independently, irrespectively of observation. The thing represented exists independently of representation, but the representation itself obviously doesn't.

The 'physical' world displayed on the screen of perception *is* representation, not the thing represented; for we associate physicality with the contents of perception. These contents of perception, in turn, are *not* the world as it actually is in and of itself, for perception is not a transparent window into reality. Therefore, we have no grounds to extend the notion of physicality beyond perception, towards the thing perceived; not any more than we have grounds to say that the clouds in the sky are 'dashboardical.' As such, we cannot say that the *real* world is physical. In fact, under any recognizable sense of the word physical the real world is *not* physical, for the very meaning of physicality is anchored in the parameters and scales of the dashboard.

To make things a bit less ambiguous, allow me to introduce a very simple convention. I shall say that something is 'physical' *in the colloquial sense* when it consists of *contents of perception;* i.e., something made of the colors, sounds, smells, flavors, or textures we *experience*. In this case, I shall use scare quotes around the qualifier 'physical.' Notice that this is precisely how we conversationally use the term when alluding to the physicality — the *felt* concreteness, solidity, coldness, texture — of, e.g., a stone in our hand. 'Physical' things are primarily experiential, in that the contents of perception are experiential. For all other usages of the word — e.g., when I refer to something pertaining the science of physics, such as physical equations or physical quantities, or when I define a particular usage in the running text — I shall use the qualifier physical *without scare quotes.* You may have noticed that I have already been using this convention in the foregoing, before making it explicit.

So let us proceed. 'Physical' states are internal cognitive states describable through *physical quantities,* such as grams and meters. This is what physical quantities were created to do: to *describe* 'physical' states. When we say, for instance, that a piece of luggage weighs 50 kilograms, we are providing a description of the perceptual experience of lifting the piece of luggage, or of seeing the behavior of a weighing instrument interacting with the luggage. If I tell someone who is about to pick up my luggage that it weighs 50 kilograms, then that person will expect a different experience than they would have expected had I said, instead, that the luggage weighs 5 kilograms. Similarly, when I tell a lost driver that their destination is 1 kilometer away, they will expect an experience of driving to their destination different from what they would have expected had I said, instead, that the destination is 100 kilometers away. *Physical quantities are relative descriptions of perceptual experience.*

Now, since 'physical' states are internal representations, the *real* states of the world out there aren't necessarily describable through physical quantities; for the characteristics of the contents of perception aren't necessarily the characteristics of the real world, just as the characteristics of dials on a dashboard aren't the characteristics of clouds in the sky. As such, we have every reason to believe that the *real* states of the world— which are external to, and independent of, our perceptual representations—are *nonphysical,* in the sense of not being amenable to description by physical quantities. After all, the latter were created to describe 'physical' states, not *real* ones.

Is it coherent to conceive of nonphysical states in this sense? Of course it is. Moreover, we all become acquainted with nonphysical states—in the sense of states that are not amenable to description by physical quantities—every single day, multiple times a day. Take a thought, for instance: what is the length, in meters, of a thought? What is the mass, in grams, of an emotion? What is the angular momentum of an intuition? The frequency, in Hertz, of an insight? Our *endogenous* experiential states—i.e., mental states that arise spontaneously within us, independently of perception, such as thoughts and emotions— are not amenable to description by physical quantities. As such, they are nonphysical states, in the sense in question. (One may feel tempted to make the theoretical and abstract assumption that they must be somehow *reducible to*—i.e., explainable in terms of—physical states, but this would be a form of circular reasoning, as the point in contention is precisely what is or isn't reducible; moreover, it would also not change the fact that the qualities of endogenous experiential states cannot, in themselves, be described in physical terms.) Yet, endogenous experiential states palpably exist: we all know them by direct acquaintance. The existence of nonphysical states is thus not an abstract conceptual inference, but a self-evident, primary empirical reality.

In saying this, I am *not* asserting that the real world, as it is in itself, is made of thoughts and emotions qualitatively akin to our own; I don't know that, for I don't know what it is like to *be* the real world out there. But our thoughts and emotions do provide proof of existence of the *kind* in question, i.e., nonphysical states. As such, it is coherent to posit that the real world is constituted of nonphysical states, whatever their specific characteristics may or may not be. And this is the scenario that cutting-edge analytic reasoning forces us to contemplate.

In summary, the contents of perception are not the world as it is in itself. Perception is not a transparent window into the world, but an encoded representation thereof, evolved to limit our internal entropy and enable effective responses to environmental challenges. What we refer to as the 'physical' world are mere representations: contents of perception amenable to description by physical quantities. But the real world out there, as it is in itself, outside perception, is constituted of nonphysical states, in the sense of states that aren't amenable to description by physical quantities. Examples of nonphysical states are our own endogenous experiences, such as our thoughts and emotions.

It will be easier to understand the next chapters if you maintain the discipline, as you read through them, to interpret what is said under the light of the discussion above. Psychologically challenging as it admittedly is, you must try to set aside—even if only for the sake of argument—the notion that you perceive the world as it actually is. A technique to do so is to imagine yourself locked up in an airplane cockpit without windows, equipped only with a dashboard of dials in the form of a virtual reality headset. The world you perceive is what is displayed inside the headset, not what is really out there. *The states of the real world lie 'behind' and 'beyond' physicality just as the sky lies behind and beyond the airplane's dashboard.* And since the only concrete examples we have of nonphysical states are endogenous experiential states, such as thoughts and emotions, physicality is most parsimoniously and coherently regarded as a dashboard representation of transpersonal mental states out there in the real world.

Chapter 3

How Physicalism gets it wrong

The physicalist starts where we all start: from our perceptions of a world surrounding us. Just like the rest of us, the physicalist perceives myriad 'physical' objects and events describable by physical quantities, occupying the scaffolding of spacetime we all seem to inhabit. And admittedly, all these 'physical' objects and events do *seem* to be external to us, in the sense that their behavior cannot be directly controlled by our thoughts, wishes, or imagination.

Naturally, internal representations of external entities *seem* to be external to us, in that their behavior—modulated as it is by the real external entities they represent—is not directly controllable by our inner mentation. This doesn't mean that our perceptual representations are external to us; it means only that the entities they represent are.

But the physicalist fails to see this distinction. Taken in by the illusion that perception is a transparent window into reality—a forgivable illusion perhaps in the 18th century, but one hard to defend among thoughtful and educated 21st-century scientists and philosophers—the physicalist mistakes the representation for the thing represented, and therefore the externality of the thing represented for the externality of the representation itself. In other words, the physicalist thinks that the 'physical' world perceived is external to the perceiver and, therefore, the *real* world out there. This is the physicalist's first error.

Don't get me wrong: the physicalist does acknowledge that the colors we see, the flavors we taste, the textures we feel, etc., insofar as they are qualities of experience, are internal cognitive representations of our own; they aren't out there in the external world. But the physicalist nonetheless believes that the *contours* discernable on the screen of perception—i.e., the outlines, forms, shapes of the contents of perception, and the relative geometrical relationships among them—are the contours of the external world. It is in this sense that the physicalist mistakes perception for the thing perceived: the forms, shapes, positions, and motions of the contents of perception are, for the physicalist, the *actual* forms, shapes, positions, and motions of the external world; the basketball perceived supposedly exists *as a ball* out there in the world, and when the basketball bounces in perception, the real ball supposedly bounces out there in the world. This is why the qualifier 'physical' has become so ambiguous in our language, referring both to contents of perception and the world as it is in itself.

Notice that the physicalist assumption here is as big and untrivial as it is unjustified, for it implies that the structure and dynamics of perceptual representations are the structure and dynamics of what is represented; that the 'physical' objects discernable in perception have a one-to-one correspondence with objects that really exist out there, with the same outlines; that the 'physical' events discernible in perception actually unfold in the external world. Why would any of this be true? A needle moving within an airplane's dashboard dial doesn't entail or imply that there is a needle moving in the sky outside (even though it does provide accurate indication of *something* happening in the sky, which doesn't need to be needle-like).

Indeed, to take the structure of the contents of perception for the structure of the external world is akin to taking the shapes of the dials on an airplane's dashboard for the shapes of the clouds, winds, and pressure distributions in the sky outside—i.e., it's just silly. There is no one-to-one correspondence between individual dials and individual objects or events in the sky, even though the objects and events in the sky are represented, in encoded form, by the dials.

Let us now discuss the physicalist's second error. At some point in our history, our culture collectively realized that *quantities* — numbers — were very useful for describing the objects and events we perceive. Such descriptions not only help us communicate — such as in telling the barman whether we want 1 or ½ pint of beer — but also model and predict the behavior of nature through scientific equations. This, in fact, is the very basis of technological development; it is sensible under any half-rational metaphysics and no sane person will deny it.

But the physicalist takes it an untrivial and peculiar step further, in what ought to be one of the weirdest moves in the history of human thought: *they argue that the description fundamentally precedes the thing described.* To the physicalist, the 50 kilograms that my piece of luggage weighs *are* my piece of luggage, not a mere description thereof; the 50 kilograms are what exists objectively out there, in external reality, giving the piece of luggage its very substance, its very essence. Indeed, the physicalist posits that the numbers with which we describe the perceived world are the *fundamental* aspect of reality; the bottom, irreducible, most solid layer of nature, existing objectively, independently of us. What *really* exists out there is not color, flavor, or scent, but meters, kilograms, seconds, liters, etc. For the physicalist, the real world is physical not merely in the sense of being describable by physical quantities, but of being *constituted* by physical quantities.

Notice the acrobatic reversal at work here: obviously it is us, human beings, who come up with our own *descriptions* of our perceptual experience of an external world. The thing that is ultimately described through its physical representation — i.e., that which is represented — is out there, while the descriptions themselves are 'in here,' so to speak. But the physicalist flips this upside down or, rather, inside out: it is only the descriptions that are out there — i.e., the kilograms, the meters, the liters, the seconds, etc. — while the things described somehow arise from, and are given substance by, the descriptions. This is literally equivalent to stating that the map precedes the territory and somehow generates the territory.

Naturally, very few people are insane enough to come up with such crowning absurdity out of the blue, without some kind of trigger or motivation, ultimately invalid as the latter may be. So how did this bizarre idea come about, before mere cultural momentum began to lend it a veneer of plausibility? I will answer this by telling you a story that is unlikely to be the whole truth, but at least illustrates a possible historico-psychological explanation for the lamentable state of our metaphysics today.

Let us go back to the early Enlightenment years, when members of our budding scientific community needed to carve out some metaphysical ground different from 'psyche,' so to avoid persecution by the Church. They were faced with a difficult conundrum for, aside from theoretical abstractions, everything is psychic, experiential, mental, spiritual. Even our perceptions of an external world surrounding us are mental, in the sense of being constituted by experiential qualities such as colors, flavors, scents, textures, etc. Everything touched by the wand of our knowledge—even if only ever so slightly— becomes *perforce* mental, for our knowledge itself resides in our mind. Anything non-mental is necessarily a conceptual abstraction of our mind, the abstraction itself residing in our mind. We are always cooped up in our own mentation. If there is anything outside it, then it must be brought inside—and thereby immediately become mental—if we are to know it in any sense of the word.

Yet, the survival demand placed on the founders of the Enlightenment was precisely to escape their own mentation— for mentation is psyche, the domain of the Church—while pursuing knowledge. How to perform such a magic trick? One possibility was to appeal to second-hand knowledge: things that are known, but not directly by ourselves. Those things, despite being known, are outside our own mind and therefore objective from our perspective. They do exist—for they are known—but without being touched by the wand of our own knowledge. However, this obviously doesn't work: known things must still exist in *a mind*, even if not our own mind; even if not a *human* mind. As such, they still fall within the scope of the psyche, the domain of the Church. Bishop Berkeley's Subjective Idealism made this abundantly clear, in claiming that the external world exists in the mind of God.

What was needed to escape the clutches of the Church was a space outside *any* mind, even the mind of God. But by looking at the world harder, deeper, all we succeed in doing is extending the scope of our own mentation; and our Enlightenment heroes understood this. *What was needed was to forfeit the concreteness of experience and reify, instead, some form of abstraction.* The claim would be that, even though the abstraction itself was undeniably mental, it could correctly *point to* something outside itself, which in turn could be said to be non-mental. This non-mental space would thus fall outside the jurisdiction of the Church.

And here is where those quantitative descriptions we talked about earlier suddenly came in handy, very handy indeed! The defining characteristic of the mental is its *qualitative* nature, i.e., what it *feels* like to know or be acquainted with something. If a semantic dichotomy could be created—invented, made up, conjured into existence—between *qualities* on the one hand, and *quantities* on the other, then our descriptions of reality could be said to have a non-mental character insofar as they are purely quantitative. And hence it was claimed that 'quality = mind' and 'quantity = non-mind.' Presto!

This magic trick is, of course, not only entirely arbitrary and merely semantic, but also self-defeating, as quantities are just mental descriptions of the mental. Nonetheless, it is reasonable to assume that it saved the lives or livelihoods of many a scientist. To this day it remains a core part of our inherited, culture-bound intuitions: just ask yourself whether you don't find it rather logical that qualities and quantities form a real dichotomy.

And so it was that the absurd notion that quantitative descriptions are fundamentally—i.e., metaphysically, ontologically—distinct from the qualitative world described became enshrined in the highest echelons of our culture. This is how magic is done: through language tricks mistaken for real facts, even—perhaps especially—by those who invent the tricks.

Originally, in the time of René Descartes, all that was needed to escape the Church was to separate the mental from the made-up non-mental—labeled the 'material'—and claim that science busied itself solely with the latter. Descartes' own substance dualism did not try to eliminate either side of this pair or make one more fundamental than the other. They were, instead, supposed to be complementary. This intellectual ethos prevailed among learned elites all the way into the early 19th century, as one can see, for instance, in these words of the great Goethe:

> Whoever can't get it into his head that mind and matter, soul and body ... were, are, and will be the necessary double ingredients of the universe, ... whoever cannot rise to the level of this idea ought to have given up thinking long ago.
> (As quoted in Rüdiger Safranski's *Goethe: Life as a Work of Art*, WW Norton, 2018, chapter 29.)

Notice that, for Goethe, some form of substance dualism was not at all a matter of faith, but one of *reason*, for failing to acknowledge it represented an abandonment of thinking itself. Goethe was an ennobled—bourgeois by birth, being the son of a financially-independent lawyer—member of the intellectual elite of his time; perhaps *the* most prominent one. His perspective is thus quite relevant and representative.

By the second half of the 19th century, however, when the game was no longer the mere survival of bourgeois intellectual elites, but their *cultural hegemony* over the clergy, an extra claim became mainstream among those elites: quantitative descriptions *precede* the qualities described, somehow giving rise and essence to the latter. This meant that what science studies—i.e., matter—is deeper and more fundamental than the Church's domain of the psyche. The equality between mind and matter was abandoned in favor of the latter. Indeed, intellectual bourgeois hero Charles Darwin—son of a successful doctor and financier—had already dealt a blow to the clergy by taking away from them the power to explain life itself. This emboldened the ambitions of the bourgeoise, so that claiming that mind must be reducible to matter was the psychologically predictable next attempt at a metaphysical *coup de grâce* against the Church. To this day, almost two centuries later, the claim is still in vigor, for modern Physicalism maintains precisely that the qualitative-mental can be reduced—in principle—to the quantitative-physical, even though nobody can even begin to explicate how that might work.

The point I am trying to make is that mainstream Physicalism is not a hypothesis motivated by evidence and clear thinking, but a philosophical side-effect of a psycho-socio-political power game. What passes for empirical evidence in favor of Physicalism is often evidence merely for the existence of a world outside our *individual* minds, not for a world metaphysically different from mind in general, as an ontological category or kind of existent. But because we are conditioned to thinking of everything outside the minds of living beings as non-mental, we naively misconstrue the undeniable and overwhelming evidence for a world outside living beings—a world that living beings *inhabit*—as evidence for the non-mental.

What leads to such interpretational bias? The answer is Physicalism itself, for it is only under its premises that mind, being supposedly a product of metabolism, must always be confined to living beings (the fact that mental states correlate well with metabolic brain states is acknowledged, but also does not imply Physicalism, as I shall discuss in detail later). Therefore—or so the thought goes—the environment *inhabited* by living beings cannot itself be mental; ergo, it is material, for what else is there? This is an example of the circularity that underpins mainstream Physicalism, as mentioned earlier. If you come from a physicalist background yourself, you may even need to reread this and the previous paragraph, perhaps a couple of times, to even see the circularity in question.

Here is another example: because everything happens *as if* what appears on the screen of perception were the real world out there, we conclude that the real world must be physical, in the sense of having the structure of the contents of perception. If one sees a train coming and then steps in front of it, one dies; it all works as though a real physical train were coming. Yet, the same applies to an airplane without windows: everything happens as if the dashboard were the sky outside; so much so that the airplane will crash if the pilot ignores the dashboard or acts against its indications. Why is that? Because the behavior of the dashboard is *correlated*, by construction, with the salient parameters of the sky outside, insofar as it *represents*—in an encoded form—the states of the sky. In other words, the dashboard conveys *accurate and important* information about the sky outside, without *being* the sky. It was built to do precisely this. But since we've become blind to the rather trivial fact that accurate information can be conveyed through representational mediation, we fail to see that perception is more akin to a dashboard than to reality. And what inculcates this bias in our minds? The answer is, again, unexamined physicalist premises, according to which the structure of the real world is 'obviously' the structure of what is displayed on the screen of perception. Physicalism is thus largely a self-perpetuating delusion. That physicalists think of themselves as being guided by evidence merely betrays the circular character of the delusion, insofar as their evidence is trivially misconstrued to imply what it doesn't.

There is more to be said about how we misconstrue evidence to acquiesce to our physicalist bias, and I shall come back to it later. For now, though, the salient point is this: while there was initial clarity among the people involved in the social power game that Physicalism was a political move, today this clarity has been lost. We now *actually* believe in Physicalism, for the cultural momentum it has amassed for being repeatedly pronounced as fact over generations—as well as its now-conditioned association with 'educated' and 'elite' perspectives—has become formidable. Just like political propagandists who eventually start gulping down their own snake oil, we now believe Physicalism wholesale, because it has been repeated *ad nauseam* by otherwise credible and educated people who were supposedly thinking about these things before we were born. This cultural momentum gives us license to *not* think critically about it ourselves, for others have done the hard thinking for us already, right? And if all those educated people believed in Physicalism, then it must be true, and we don't need to spend our energy reinventing the wheel ... right? All we need to do, when it comes to our own credibility, career, and social standing, is to repeat the physicalist story ourselves, so we also come across to others as educated elite thinkers, who courageously—even heroically—face the tough fact that nature is dead and meaningless. And thus, this self-perpetuating self-deception endures robustly, for your children watch you—or the evening news anchor, the family doctor, the teacher at school, etc.—gulp down the snake oil. They then repeat it to your grandchildren, and these in turn to your great-grandchildren, etc. After generations of this pernicious psychological cycle, a culture can *sincerely* become quite certain that in-your-face balderdash—such as the map preceding the territory—holds water, for a manufactured sense of plausibility eventually saturates it. Welcome to the 20th century, a time when the Church was already largely defeated, but bourgeois intellectual elites stepped on their own mines while returning from the battlefield.

Fortunately, sociopsychological dynamics cannot make something incoherent and empirically inadequate magically become true. There is an indelibility to reason and evidence that, like water splashing against rock, eventually cracks the strongest psychological walls. And so this, too, began to happen in the late 20th century.

Under mainstream Physicalism, physical entities are *defined* in terms of their measurable, *quantitative* physical properties. In other words, an electron *is* its measurable mass, charge, momentum, etc.; there is supposedly nothing to an electron but its quantitative properties. Still under Physicalism, these physical entities have *standalone existence*: they supposedly exist in and of themselves, independently of observation or measurement. Observation and measurement merely *disclose, unveil* their properties, which already existed—or so the story goes—immediately prior to the observation or measurement. This notion is called 'physical realism.' And sure enough, the practice of empirical science did not contradict it up until the late 1970s.

But then, while looking closer and closer into the primary building blocks of matter, scientists noticed something that defied physicalist expectations: as it turns out, laboratory results began to show that physical entities in fact can*not* be said to exist prior to measurement. Instead, *physicality is a product of measurement.*

This remarkable series of experiments—refined by different research groups over the span of more than four decades—moved the Nobel Prize committee to award the lead investigators the Nobel Prize in physics in 2022; the highest honor in science. I shall now briefly describe the general form of the experiments and discuss why their results refute physical realism.

The experimental procedure goes as follows: two subatomic particles—say, A and B—are prepared together, so that they are entangled ('entanglement' is physics jargon for saying that the particles cannot be described independently of one another). They are then shot in opposite directions at (near) the speed of light. After a certain distance is covered, a first scientist—let's say, Alice—measures particle A, while another scientist—say, Bob—simultaneously measures particle B at a different, far away location. What then transpires is that Alice's choice of what to measure about particle A determines what Bob sees when he observes particle B. Let me repeat this, so it sinks in: *what one scientist chooses to measure about one particle determines what the other scientist sees when he looks at the other particle.*

How can this be? How can the choice of what to measure about one particle determine what the other particle *is*? Shouldn't observation merely *reveal* what the particle already was, in and of itself, regardless of what is measured about another? And how can two distant but concurrent measurements be entirely correlated with one another, despite the speed-of-light limit that should preclude any information transfer required for such correlation?

This result is not reconcilable with physicalist premises (unless some grotesque science fiction fantasies are taken seriously, which I shall discuss shortly). If the two particles were *physically real*, in the sense of having standalone existence, then their measurable properties—which, as we've seen, define their existence—would be whatever they are regardless of what one chooses to measure about them. Take a piece of luggage, for instance: it seems to have a certain mass, height, and length regardless of what is being measured about it. If it weighs 50 kilograms sitting on a weighing scale, then it will still weigh 50 kilograms even when it's not sitting on a weighing scale— or so Physicalism stipulates. Measurements supposedly *reveal* something that was already the case about the piece of luggage immediately before the measurement was done, not determine it. If mainstream Physicalism were true, the same should apply to the subatomic particles in our experiment, for a piece of luggage is simply a compound aggregation of subatomic particles: measurement should simply reveal the properties the particles already had, in and of themselves, immediately prior to the measurement.

Experimentally, however, what we see is that the properties of one particle depend on what we choose to observe about the other. The particles' properties don't have standalone existence but are, instead, created by the very act of measurement. And since there is nothing about a physical particle but its measurable physical properties, the *particles themselves* cannot be said to exist unless and until a measurement is done. This, of course, is incompatible with physical realism and, therefore, mainstream Physicalism itself.

If one still insists on holding on to physical realism, one has to part with explicit, level-headed, empirically based science; one has, instead, to entertain one of two highly inflationary and entirely speculative fantasies. The first is the so-called 'Everettian Many-Worlds' hypothesis: every time an observation is made, *all* possible outcomes are supposed to be produced, but each in a separate, parallel universe. The paradigm-defying outcome we *happen* to see is the one that *happens* to be produced in the parallel universe we *happen* to inhabit. Copies of us in other parallel universes observe all the other outcomes, so there is nothing to fret about if we see stuff that contradicts our expectations and prejudices. One can almost feel the warm, fuzzy metaphysical reassurance this provides: whatever you choose to believe is sort of fair game, for everything that *could* be observed *is* observed, just in some other *inaccessible* universe, in some other *inaccessible* dimension, by some other *inaccessible* copy of you; 'inaccessible' being the operative word here. This undisguised but admittedly very imaginative subterfuge, if taken seriously, could be used to justify just about anything that doesn't outright contradict laws of large numbers.

Some claim that the idea of parallel universes 'flows naturally' from quantum theory, a notion grounded in a combination of grotesque epistemic arrogance and a complete abandonment of one's natural sense of plausibility. The idea is that, because the equations of quantum mechanics — which *we* have come up with to try and get a handle on nature's behavior — cannot predict the outcome of any specific event, but only statistical averages, then of course nature must produce *all possible* outcomes; otherwise we, godly intellects that we are, certainly would have been able to do better by now, wouldn't we? In other words, because *we*, bipedal apes, haven't managed to predict nature's behavior at its finest-grained level, then ... invisible parallel universes!

This idea would be a little more apt if we had some, *any*, direct evidence for all this parallel stuff. Alas, we don't. We must simply believe in countless parallel universes popping into existence every infinitesimal fraction of a second — every time there is a microscopic interaction anywhere in the universe — which is arguably the most inflationary notion that human thought *can* coherently produce. It is so inflationary, in fact, that it is literally *impossible* to explicitly visualize how much stuff popping into existence is entailed by it; it's just too much to wrap one's head around; it's a dizzying, exponential, thermonuclear explosion of empirically unverifiable stuff that makes the Big Bang look like a bang snap. That such fantasy is not only taken seriously, but even publicly promoted by professors from some respected universities, illustrates how far belief in Physicalism can take otherwise reasonable, intelligent people down extraordinarily implausible avenues of pure speculation. From the perspective of psychology, this is deserving of in-depth study, and I don't say this sarcastically at all.

The other entirely speculative and supremely vague fantasy is called 'superdeterminism': there supposedly are mysterious 'hidden variables' in nature—emphasis on 'mysterious' and 'hidden'—that do exactly whatever needs to be done for the experimental results obtained to remain consistent with physical realism. What are these hidden variables? No one has ever specified them explicitly and coherently, so we can't even start looking for them through experiments that could falsify the hypothesis. How, precisely, do the hidden variables do what they are presumed to do? No one has ever specified that either; they just somehow do it. But do what, exactly? Whatever must happen in nature so we can continue to believe in physical realism, despite experimental results telling us otherwise. If there is any exaggeration in this colloquial characterization of superdeterminism, it is only mild.

Superdeterminism is akin to saying, if you believe in Creationism, that nature has a mysterious, hidden agent who does exactly whatever needs to be done to create the illusion of a fossil record, even though natural selection is false and the world was created within the past ten thousand years. How? We have no idea. What is this mysterious agent? We have no idea; we just call it 'hidden god.' And we define this agent in terms of whatever needs to be true so to enable us to continue to believe in Creationism, despite overwhelming empirical evidence for an alternative hypothesis—namely, that the fossil record points to evolution by natural selection. Underneath its highly technical language, the spirit of superdeterminism is surprisingly akin to this.

You see, what the laboratory results have been consistently telling us for over 40 years—so consistently, in fact, that a Nobel Prize has been awarded to the investigators—is that physical entities have no standalone existence. They are, instead, products of measurement. But since this result is metaphysically unacceptable to some, they conjure up undefined hidden variables and inaccessible parallel universes to rescue our metaphysical prejudices from the cold clutches of hard experimental evidence.

To impress upon you the fact that I am not exaggerating, we can look a little deeper into superdeterminism. According to it, the settings of the measurement devices used by Alice and Bob somehow change what the particles A and B *are*—as opposed to simply, well, *measuring* them, which is what measurement devices are made to do—thereby creating the correlations between what Alice and Bob see. This is akin to saying that, when you photograph the moon up in the night sky, the aperture and exposure settings of your camera change what the moon *is*. This way, regardless of what you see in the resulting photograph, you don't have to part with your favorite theoretical prejudice about the nature of the moon, for the moon that was there just before you photographed it was different from the moon on the photo. Moreover, Alice's and Bob's measurement devices have to somehow conspire with each other, instantaneously and at a distance, so to ensure that the physical properties of particles A and B are correlated, despite speed-of-light constraints. But this, too, is miraculously taken care of by the unspecified hidden variables, whatever they may be.

By reifying Physicalism to the position of *necessary, a priori truth*, despite evidence to the contrary, our culture has lent legitimacy to fantasies that are beyond implausible. After all, since Physicalism *must* be true, any way to reconcile evidence with it, no matter how desperate and implausible, must be a legitimate part of the debate, right? And so our rational intuitions of plausibility are thrown unceremoniously out the window. This is how cultures lose themselves to their own nonsense.

Short of theoretical fantasies, we must thus accept, on hard empirical grounds, that the physical world is created upon observation or measurement. In other words, physics is telling us experimentally that, just as we've concluded before on entirely different grounds, the physical world is but a dashboard representation created by measurement, not the real world out there. *The only physical world there is is the 'physical' world on the screen of perception*; there is no underlying, purely quantitative, abstract physical world with standalone existence.

None of these experimental results is actually surprising or discombobulating when regarded without metaphysical prejudice: the dials in an airplane's dashboard only show something when a measurement is made, for what they show is precisely the outcome of the measurement. Without a measurement, the needles in the dials don't move and nothing is shown, for there is nothing to be shown. Is this difficult to understand? Now, in precisely the same way, as experiments have repeatedly indicated, physical entities are dashboard representations of measurement outcomes, so that without measurement no physical entities can exist. Is *that* difficult to understand? If no measurement is performed, the dials have nothing to show and, therefore, there is no physical world; for the physical world is *constituted by* the dial indications. In other words, all physical entities are merely 'physical' entities.

None of this implies that there is no reality prior to measurement, otherwise we would have an even bigger problem, as there would be nothing to be measured in the first place. But there is still a sky when no airplanes are flying around making measurements. Without airplane sensors, there just aren't any *dashboard representations* of the sky. But the sky itself—unmeasured—is still there. In exactly the same way, when we don't measure the real, external world, there is no *'physical'* world; for the 'physical' world—as displayed on the screen of perception—is but an internal representation of our measurements of the real world. Nonetheless, the real, nonphysical world is still there, regardless.

Things couldn't be simpler if we just accepted what nature is telling us, as opposed to forcing our metaphysical prejudices upon nature: physicality is not the real world, but an internal cognitive representation thereof; that's why it only appears upon observation and can't be said to exist prior to observation. The world that is measured, in turn, is *real*, but not physical, in the sense of not being describable through physical quantities. That's all there is to it, and it isn't difficult to understand.

The dashboard metaphor can even make straightforward sense of the instantaneous correlations between what Alice and Bob see upon measuring particles A and B, respectively. These correlations are only puzzling if we assume that the particles have standalone existence, but not if they are mere representations. To see this, consider the following analogy.

Imagine that you are watching a football match at home. Because you are such a great fan of football, you bought two large TVs to follow the same match, simultaneously, on two different channels. Imagine also that the two different broadcasters have their own cameras in the stadium, so each channel shows *different* images of *the same* match. And you watch the two different images side by side.

Now, obviously, the two images will be entirely correlated with one another, for they are representations of the same match, the same underlying reality. The images have no standalone existence, only the football match in the stadium—the thing in itself—has. Nonetheless, the images will also be different, for they are produced by different cameras and camera angles. None of this is counterintuitive or difficult to understand.

However, if you were a time traveler from the 18th century and didn't understand how TVs work, you would be flabbergasted by the correlations between the two images: how can the little men running inside the box to the left move in perfect synchronization with the other little men running inside the box to the right? How can that happen even when the boxes are totally isolated from one another, so that the little men can't talk to each other across the boxes? Incomprehensible!

Of course, the source of this puzzlement is the unexamined assumption, by our time traveler, that the images aren't mere representations, *but the things in themselves.* If you think that there are *real* little men, with standalone existence, running inside the two TV sets, the correlation of their behavior across the sets would seem magical indeed. And this is precisely the mistake we make when it comes to the laboratory experiments being discussed here: we think of the entangled particles A and B as *real* things in themselves, not mere representations of an underlying nonphysical reality. If we understood and accepted the latter, the experiments wouldn't seem magical at all: the entangled particles are two different representations— two different images, two different camera angles—of the same underlying reality; that's why they are correlated instantaneously and at a distance, just like the images on the two TV sets are instantaneously correlated at a distance. But instead of acknowledging what nature is telling us, we insist on thinking like 18th-century people in the face of 21st-century experimental evidence.

Quantum physics experiments are not the only instance in which laboratory results directly contradict physicalist premises and expectations. Since 2012, results in the field of neuroscience of consciousness have been doing the same, with overwhelming consistency. For instance, before 2012 the generally accepted wisdom was that psychedelic substances, which lead to unfathomably rich experiential states, did so by stimulating neuronal activity and lighting up the brain like a Christmas tree. Modern neuroimaging, however, now shows that they do precisely the opposite: the foremost physiological effect of psychedelics in the brain is to significantly reduce activity in multiple brain areas, while increasing it nowhere in the brain beyond measurement error. This has been consistently demonstrated for multiple psychedelic substances (psilocybin, LSD, DMT), with the use of multiple neuroimaging technologies (EEG, MEG, fMRI), and by a variety of different research groups (in Switzerland, Brazil, the United Kingdom, etc.). Neuroscientist Prof. Edward F. Kelly and I published an essay on *Scientific American*'s website (titled "Misreporting and Confirmation Bias in Psychedelic Research," on 3 September 2018) providing an overview of, and references to, many of these studies. As Prof. Kelly put it, "impressive and direct measurements of decreased brain activity" are by far the most robust effect that psychedelics have on the brain.

This result contradicts mainstream Physicalism for obvious reasons: experience is supposed to be generated by metabolic neuronal activity. A dead person with no metabolism experiences nothing because their brain has no activity. A living person does because their brain does have metabolic activity — or so the story goes. And since neuronal activity supposedly *causes* experiences, there can be nothing to experience but what can be traced back to patterns of neuronal activity (otherwise, one would have to speak of disembodied experience). Ergo, richer, more intense experience — such as the psychedelic state — should be accompanied by increased activity *somewhere* in the brain; for it is this increase that supposedly causes the increased richness and intensity of the experience (this rationale applies even under the understanding that experience correlates with intrinsic information, provided that more than half of the associated neurons remain inactive in the psychedelic state, which is the case).

Notice that Physicalism would remain consistent with an *overall* decrease of brain activity in the psychedelic state, provided that one could still find *localized* increases in parts of the brain consistent with the experience. The reason for this is that, under Physicalism, not all neuronal processes lead to experience; only the so-called 'Neural Correlates of Consciousness' (NCCs) supposedly do. It is thus conceivable that psychedelics could reduce activity in processes not related to conscious experience, while leading to localized increases in the NCCs. In particular, it is conceivable that psychedelics could impair inhibitory processes that, once impaired, disinhibit the NCCs. The problem is that all this relies on there being plausibly sufficient increases of activity *somewhere* in the brain—corresponding to the now-disinhibited NCCs—compared to the baseline, so as to account for the increase in the richness and intensity of experience. But no such a thing has been seen.

Since brain activity doesn't increase in the psychedelic state, physicalist neuroscientists then conclude that something else in the brain must. And so the hunt is on for *something* in the brain that increases under the effect of psychedelics. Many possibilities have been proposed and somewhat fallen by the wayside, such as brain activity variability and functional connectivity. But one remains and is significantly hyped as the best physicalist hypothesis for accounting for the psychedelic experience. It goes by various names, such as 'brain entropy,' 'complexity,' 'diversity,' and so on (see "The entropic brain – revisited," by Robin Carhart-Harris, published in *Neuropharmacology*, 2018). But what it means is very straightforward: brain *noise*—i.e., residual brain activity that unfolds according to no discernible pattern; brain 'TV static,' if you like.

The idea here is that, although brain activity decreases with psychedelics, the residual activity that remains is desynchronized by the drug, thereby becoming relatively more random than in the baseline. And this relative increase in randomness or entropy—the latter meaning the degree of disorder of the remaining brain activity—is supposed to account for the unfathomable experiential immensity of the psychedelic state. The logic is that more random activity contains more *Information* than synchronized activity with discernible patterns. Under a certain definition of 'Information,' which I shall elucidate below, this is indeed true. And thus, the *extra Information* physiologically imparted by psychedelics supposedly accounts for the *extra richness and intensity* of the psychedelic experience.

There are many reasons why this 'entropic brain hypothesis' is implausible to the point of being ludicrous, so let's tackle them systematically, starting with the underlying logic discussed above. The fallacy of trying to account for richer, more intense experience in terms of higher Information content is that it relies on conflating two completely different definitions of the word 'information.'

The first definition was that coined by Claude Shannon, father of information theory, in his seminal 1948 paper, "A Mathematical Theory of Communication." The idea there is that Information is a measure of the level of 'surprise' embedded in a message or signal. More specifically, the more alternative possibilities are eliminated by a message or signal, the more 'surprise value'—and, therefore, Information—it contains. For example, if a message stated simply that a certain person is married, then only one other possibility would be eliminated: namely, that the person is single. The level of 'surprise' here is only 50%, since only one out of two possibilities can be eliminated by the message. But if a message were to contain, say, a picture of the cloud cover over your city, countless other possible patterns of cloud cover would be eliminated by it, and the level of 'surprise' would be much greater. That picture would thus contain a lot more Information.

One way to operationalize this particular definition of Information is to think in terms of compression. A photograph — playing the role of message, or signal — with clear and repeated visual patterns is compressible and can, therefore, be stored in a smaller computer file. The discernible patterns allow the compression algorithm to discard many pixels from the original image, since the algorithm can later reconstruct those pixels based on knowledge of the patterns according to which they appeared in the first place. For example, a photograph of an empty chessboard is highly compressible, because the black and white pixels appear on it according to a very regular pattern, so there is no need to store each and every pixel; all we need is to know the pattern of a chessboard. But a photograph of TV static is much less compressible, for the black and white pixels do not follow any recognizable pattern. In this latter case, nearly all pixels need to be stored.

Shannon's definition of Information means that, the more compressible a signal is, the less Information it has, for knowledge of the associated patterns reduces the degree of 'surprise' we have when we analyze the signal. By the same token, the less compressible a signal is, the more Information it contains, for our inability to recognize underlying patterns renders many 'pixels' in it unexpected and, therefore, 'surprising.' When I use the word 'Information' in Shannon's sense, I capitalize it, as I have already been doing.

Now, Shannon's definition of 'Information' is a very technical one, invented for very specific purposes in communications engineering: namely, to calculate the minimal bandwidth of the communication channel required to transmit the message after compression. It doesn't — and was never meant to — replace the *colloquial* use of the word. In the colloquial sense, the word 'information' (this time *not* capitalized) means the amount of *semantic content* of a message or signal. This way, a message or signal has a lot of information if it *means* a lot. On the other hand, a message that means nothing has *no* information.

The crucial thing to notice here is that, in a very important sense, Information and information are *opposites*. A completely random and uncompressible signal has maximum Information, but no information; for a random signal *means nothing*: it has no discernible structures or patterns that could be recognized and therefore unlock cognitive associations. TV static has near-maximum Shannon Information, but it means nothing. Therefore, it has no information in the colloquial sense, this being the reason why we don't sit in the living room to watch TV static; instead, we watch TV programs, which have a lot of recognizable—and, therefore, compressible—patterns in the form of objects, people, and events. As such, a signal with a lot of information has, by definition, lots of recognizable patterns, therein residing its meaning. Yet—and precisely for this reason—it has relatively little Information in Shannon's sense.

When claiming that psychedelics increase the amount of Information in the brain, the proponents of the 'entropic brain hypothesis' are using Shannon's technical definition of Information. But when claiming that an increase in the information content of the brain accounts for the richness and intensity of the psychedelic experience, they can only be appealing to the colloquial definition of information. Alas, these two denotations not only aren't the same, they effectively are opposites. The proponents' conflation of the different meanings of the word 'information' renders their entire logic nonsensical. They seem to stick to the mere word without understanding what it means in different contexts. The intuitive appeal of their hypothesis is thus no more than a linguistic phantasm.

Indeed, Shannon's Information was defined for the purpose of *communications*, as made clear in the very title of his seminal paper. It is only when we are dealing with communications that we want to know how compressible a signal is—i.e., how much Information it has—so to evaluate the minimal bandwidth of the communication channel required to transmit said signal. But when it comes to brain activity, nothing is being communicated; nothing is being transmitted through a channel; the activity already arises where it needs to be. So to apply Shannon's definition of Information here is clearly inappropriate, at best naïve, and surely misleading.

Moreover, when a subject describes a psychedelic experience as rich and intense, what the subject means is that the experience has a lot of *semantic content*; i.e., it *means* a lot to the subject, unlocking many associative links in a cognitive chain reaction. This richness of meaning is evoked by *recognizable cognitive structures and patterns*, which is the opposite of entropy. After all, a psychedelic experience isn't random or unstructured; it isn't akin to TV static. If it were, it precisely *wouldn't* be described as rich or intense, but mind-numbingly boring instead; for there is nothing more devoid of evocative semantic content than TV static. A psychedelic 'trip' is so unfathomably rich and intense precisely because it has relatively little Shannon Information, and a whole lot of information in the colloquial sense. Random, entropic brain activity is thus precisely the *opposite* of what one would expect under physicalist premises; provided, of course, that one actually understands information theory. Just about anything else would be less implausible a physicalist account of the psychedelic experience.

I published this criticism of the Entropic Brain Hypothesis (EBH) on the website of the Institute of Art and Ideas (IAI), on the 21st of June 2023, under the title "Brain noise doesn't explain consciousness: A psychedelic experience isn't akin to TV static." On the 30th of June 2023, Prof. David Nutt—the most senior member of the team that originally proposed the EBH—replied in the same venue under the title "David Nutt: entropy explains consciousness: We don't need mysticism to explain psychedelic experience." The most conspicuous fact about his answer is that, despite the title chosen by the IAI, *Prof. Nutt didn't seem to even try to defend the EBH from my criticism*, opting, instead, to point to other fuzzier and even less empirically substantiated physicalist accounts of the psychedelic experience (allow me to ignore his allusion to mysticism, for it is not deserving of commentary). As I shall discuss in the next chapter, this constant switching to other vague accounts, every time one particular account is substantially criticized, renders Physicalism impossible to pin down and, therefore, meaningless. Be that as it may, it appears that even its very creators aren't prepared to explicitly defend the EBH from the criticism above, which I suppose is telling.

But even if we ignore this entire point and pretend, for the sake of argument, that Information and information are the same thing, the EBH still has no legs for obvious other reasons. I've discussed this *ad nauseam* in previous writings, so I shall limit myself to a mere summary here.

Decades of research in the neuroscience of consciousness have demonstrated consistent correlations between patterns of brain *activity* and reported inner experience. Under Physicalism, this suggests that the only plausible account of experience is brain activity. But if the EBH were correct, it would imply that, in the case of psychedelics alone, something totally else must account for experience. What is the likelihood that there are *two completely different* brain mechanisms that generate experience under physicalist premises? One cannot defend Physicalism by proposing a completely different theory of consciousness for each different set of data, as this would be grotesquely inflationary and render the scientific implications of Physicalism unfalsifiable to the point of being meaningless.

Moreover, the increase in brain *noise*—pompously called 'complexity' and 'diversity' by the proponents, which misleads casual readers into concluding that psychedelics induce more 'complex' or 'diverse' brain activity, in the colloquial sense, while the very opposite is the case—measured during the psychedelic state is ludicrously minute: it averages at 0.005 in a scale that runs from 0 to 100! (See the paper "Increased spontaneous MEG signal diversity for psychoactive doses of ketamine, LSD and psilocybin," by Michael M. Schartner *et al.*, published on 19 April 2017 in *Scientific Reports*.) The proponents' defense here is that, minute as it is, the effect is still statistically significant. But this misses the point entirely: statistical significance—an arbitrary threshold as it is—only means that the effect probably isn't a measurement or methodological artifact; it says precisely nothing about the *strength* of the effect. And the strength of the effect is key, for the proponents are trying to account for the mind-boggling richness and intensity of the psychedelic experience—a very, *very* large subjective effect—in terms of a ludicrously minute physiological effect. This stretches plausibility.

Indeed, it seems silly to suggest that a 0.005% increase in brain *noise*, of all things, accounts for life-changing, imaginary trips to other dimensions, conversations with aliens, insights into the fabric of reality, the nature of the self, life, the universe, and everything. Imaginary and illusory as they may be, these experiences are real *as such*; i.e., *as experiences*. And so they must be accounted for, under Physicalism, in terms of plausibly comparable physiological effects. Short of an appeal to magic, a minute increase in brain noise just isn't one such a physiological effect.

Finally, the brain noise increase observed during the psychedelic state is a statistical average. For some of the placebo-drug pairs studied, brain noise went the other way: it *decreased*. Nonetheless, those subjects also had the psychedelic experience. What, then, accounts for *their* experiences under physicalist premises?

Analytic Idealism accommodates all these empirical observations without the problems and contradictions entailed by mainstream Physicalism. Under Idealism, the brain and its patterns of neuronal activity are not the *cause* of inner experience, but the *image*, the *extrinsic appearance* of inner experience. In other words, brain activity is what inner experience *looks like* when observed from the outside. As such, the correlations ordinarily observed between patterns of brain activity and inner experience are due to the trivial fact that the appearance of a phenomenon correlates with the phenomenon. And the break of this correlation observed in the psychedelic state is due to the fact that, unlike a *cause*, the *appearance* of a phenomenon doesn't need to be always *complete*—i.e., the appearance of a phenomenon doesn't need to always reveal everything salient about the phenomenon. For instance, if you were standing right in front of me, facing me, your appearance would not reveal your back, what's under your skin, the microscopic details of your metabolism, etc.; a great many things would remain hidden, undisclosed by your extrinsic appearance. In just the same way, the appearance of inner experience that we refer to as brain activity isn't always complete: during the psychedelic state, it leaves out quite a bit about the phenomenon it is an appearance of. And there is nothing extraordinary or counterintuitive about it.

The EBH is a linguistic charade; it is both illogical and empirically inadequate. And because the other physicalist alternatives are even fuzzier and less empirically substantiated, this leaves mainstream Physicalism unsupported as a viable metaphysics of mind. Its inability to account for verified, robust laboratory results in both physics and neuroscience betrays obvious errors on the part of the physicalist.

Yet another error the physicalist makes is to accept a fundamental and gargantuan gap in explanatory power, at the heart of Physicalism, as if it were a mere detail to be sorted out later, as opposed to an insoluble epistemic contradiction. For indeed, Physicalism fails to account not only for the psychedelic experience, but for *any* experience. And since experience is all we ultimately know — everything else being theoretical abstractions of Physicalism itself—there is an important sense in which Physicalism fails to account for *all that is known*. It explains no pre-theoretical fact of nature, only the abstractions created by its own premises, in a circular, question-begging manner.

To see why Physicalism fails to explain experience, notice that there is nothing about physical parameters—i.e., quantities and their abstract relationships, as given by, e.g., mathematical equations—in terms of which we could deduce, *in principle*, the qualities of experience. Even if neuroscientists knew, in all minute detail, the topology, network structure, electrical firing charges and timings, etc., of my visual cortex, they would still be unable to deduce, *in principle*, the experiential qualities of what I am seeing. This is the so-called 'hard problem of consciousness' that is much talked about in philosophy.

It is important to understand the claim here correctly. We know, empirically, of many correlations between measurable patterns of brain activity and inner experience. It is thus fair to say that, in many situations, we can correctly guess what experience the subject is having based solely on the subject's measured patterns of brain activity. We have even been able to tell what subjects are dreaming of just by reading out the subject's brain states. However, these correlations are *purely empirical*; that is, we don't know *why* or *how* certain specific patterns of brain activity correlate with certain specific inner experiences; we just know *that* they do, as a brute empirical fact. And if we look at enough of these brute facts, we will eventually be able to extrapolate and start making good guesses about what people are experiencing, based on their measured brain states alone. None of this implies any *understanding* or *account* of what is going on; of *how* nature allegedly goes from quantitative brain states to qualitative experiential states. These brute facts are just empirical *observations*, not explanations of anything. We don't owe brute facts to *any* theory or metaphysics, since they are observations, not accounts. Physicalism gets no credit for brute facts.

This is not just an abstract theoretical point I am trying to make here, but a very concrete one. We may know empirically that brain activity pattern, say, P_1 correlates with inner experience X_1, but we don't know *why* X_1 comes paired with P_1 instead of P_2, or P_3, P_4, $P_{whatever}$. For any *specific* experience X_n — say, the experience of tasting strawberry — we have no way to deduce what brain activity pattern P_n should be associated with it, *unless we have already empirically observed that association before*, and thus know it merely as a brute fact. This means that there is nothing about P_n in terms of which we could deduce X_n *in principle*, under physicalist premises. This is the hard problem of consciousness, and it is, in and of itself, a fatal blow to mainstream Physicalism. It means that Physicalism cannot account for *any one* experience and, therefore, for *nothing* in the domain of human knowledge.

Notice that the hard problem is a *fundamental epistemic problem*, not a merely operational or contingent one; it isn't amenable to solution with further exploration and analysis. Fundamentally, there is nothing about *quantities* in terms of which we could deduce *qualities* in principle. There is no logical bridge between X millimeters, Y grams, or Z milliseconds on the one hand, and the sweetness of strawberry, the bitterness of disappointment, or the warmth of love on the other; one can't logically derive the latter from the former.

Going the other way around, from qualities to quantities, is possible *by construction*, for quantities were invented precisely as *relative descriptions of qualities*; i.e., descriptions of the *experiential difference* between, e.g., carrying a 50Kg-heavy piece of luggage and a 5Kg-heavy one (the experiential difference is described as 45Kg); driving a car for 100Km and 1Km (the experiential difference is described as 99Km); seeing blue and seeing red (the experiential difference is described as 750THz – 430THz = 320THz).

But *the meaning of these relative descriptions is anchored in the very qualities they describe,* which thus constitute their semantic reference. In other words, the meaning of '430THz' *is* the felt quality of seeing red; the meaning of '5Kg' *is* the felt quality of lifting a 5Kg weight (or the felt quality of seeing a 5Kg weight fall within a viscous fluid, bounce off an elastic surface, lie on a weighing scale and move its needle, or whatever other experience is describable by 5Kg). As such, one cannot start from quantities and try to generate qualities from them, for in this case the semantic reference—i.e., the qualities—is supposed to *result from* the quantities, and therefore can no longer preexist them. This robs the quantities of their meaning and makes it impossible to deduce anything from them.

Let me try to clarify this with a metaphor. Trying to deduce qualities from quantities alone is like trying to pull the territory out of the map. The lines on a map only have meaning insofar as they point to a territory that preexists the map, and to which the map refers. But if we try to account for the territory in terms of the map, then the territory can no longer preexist the map— for it's now supposed to somehow arise from the map—and, therefore, the lines on the map lose their meaning entirely; nothing can be deduced from them anymore (that you could make this deduction based on other map-territory pairs you've seen before violates the spirit of the analogy; you must, instead, ask yourself whether you could deduce a territory from a map if the map were the *first and only thing* you had ever cognized in your life). This is exactly what the physicalist does when attempting to explain experiential qualities (the territory) in terms of physical quantities (the map).

The fundamental absence of a logical bridge to connect quantities to qualities, caused by the abandonment of the semantic reference that underpinned the meaning of the quantities to begin with, *is* the hard problem. The premises of mainstream Physicalism are such that, in order for quantities to have meaning, qualities need to preexist them. But when Physicalism then tries to account for the qualities in terms of the quantities, the latter cannot preexist the former anymore, and thus become literally meaningless. Nothing can be deduced in principle from meaningless things, and that's the hard problem right there.

In trying to account for the territory in terms of the map, physicalists rob the map of its meaning and become confused when they fail to explain any experience in terms of it. They then promise that one day, when new and more advanced editions of the map are developed, our descendants will be able to reach into the map and pull the territory out of it; they mistake a *fundamental epistemic contradiction* for an operational or contingent problem.

Let me try to formulate this epistemic contradiction at the heart of mainstream Physicalism in a different way. Ultimately, I will be talking about the same problem, just using different language. Yet, sometimes doing so helps one grasp the issue with more depth and nuance, especially if one understands why the two formulations are in fact the same thing.

A first physicalist premise is that, at its foundational level, reality is non-mental; i.e., physical in the sense of being purely quantitative. But since the world we perceive around us is made of mental qualities—colors, flavors, scents, etc.—the notion of a non-mental reality is necessarily a conceptual *abstraction of mind*. Whether the abstraction is correct or not—i.e., whether it correctly points to real non-mental entities or not—is epistemically irrelevant, in that the abstraction itself is created by, and always resides in, conceptual mentation.

The next premise of Physicalism is that mind is itself caused, or generated, by this conceptually abstracted non-mental substrate. Epistemically, this means that Physicalism tries to account for mind in terms of something that can only ever be known *as an abstraction of mind.* And so the physicalist ends up like a dog chasing its own tail: physicality can only be known as a conceptual abstraction, therefore mind must preexist it for it to have any meaning; but then the physicalist tries to account for mind in terms of physicality, therefore the latter must preexist mind. Alas, one can't have it both ways.

The hard problem is not at all a real problem out there in reality, a problem to be solved. Instead, it's merely an internal epistemic contradiction in the alarmingly confused mind of the physicalist. And that's all there is to it.

Here is yet another metaphor to try to clarify this. Think of a painter painting a self-portrait. Now imagine that, once the portrait is complete, the painter points triumphantly at it and declares: "I am the portrait!" At that point, the painter will face the dilemma of having to account for their own existence in terms of a pattern of pigment distribution on canvas; which is of course impossible: it was the painter who applied the pigment to the canvas in the first place, so the painter must preexist the pattern. But if so, then the pattern can't account for the painter.

The physicalist's mind does something analogous: it invents—through conceptual abstraction—the notion of something non-mental; i.e., matter, physicality in the strictly quantitative sense. Then that same mind tries to *account for itself in terms of its own invention,* just as the painter tried to account for themselves in terms of their painting. Physicalism doesn't work for the same reason that our painter's dilemma is insoluble: there is an internal epistemic contradiction in both systems of thought, which arises from self-reference, or dog-tail-chasing. The physicalist tries to account for the existence of mind through a system of signs that can only have meaning if mind preexists the signs.

So here are the errors of mainstream Physicalism that we've discussed: (a) in naively assuming that the structure of the contents of perception is the structure of reality, Physicalism mistakes the internal dashboard for the real world outside; (b) in incoherently postulating that descriptions precede the thing described, Physicalism hopelessly tries to pull the territory out of the map; (c) in stubbornly insisting that physical entities have standalone existence, Physicalism contradicts over 40 years of repeated experimental results and indulges in fantasies that should have no place in a level-headed dialogue based on reason and evidence; and (d) in failing to see that its inability to account for experience is a fundamental, internal epistemic contradiction at the heart of Physicalism, it foolishly resorts to untenable promises of future progress.

Chapter 4

How does Physicalism survive?

In Chapter 3, we've discussed the multiple errors in the thought-process underlying mainstream Physicalism. Any one of those errors, in and of itself, is sufficient to refute the physicalist hypothesis. All of them together means that Physicalism is the worst metaphysical hypothesis still seriously considered today. Which, of course, raises the question: how can something so wrong have been so popular among learned elites for so long?

We already discussed the main reasons: the social, political, and psychological payoffs of Physicalism at its inception, and then up until the late 19th century, have accrued so much cultural momentum that we are still feeling their effects today. Physicalism is like a heavy train that is very difficult to stop on its tracks, even after its engine has already cut off. Moreover, even today there are still social and psychological dynamics that maintain some of that momentum. The most obvious one is that being a physicalist helps your career in science, philosophy, and business, and gives you the aura of a courageous and mature adult who stares the difficult facts—a dead and meaningless universe—in the face. This is obvious enough and requires no further commentary. But some of the dynamics Physicalism still has going for it are much more subtle, interesting, and sometimes even hard to believe.

It is surprising, for instance, how many educated physicalists—many of whom are members of academia and hold high degrees in science and philosophy—think that Physicalism is the only metaphysical hypothesis consistent with *realism* (i.e., the notion that there is an external world independent of our individual minds, whether it's physical or otherwise), *naturalism* (i.e., the notion that nature unfolds spontaneously, according to its own intrinsic dispositions, as opposed to the whimsical interventions of a deity outside nature), *rationalism* (i.e., the notion that the human intellect can recognize and model the regularities of nature's behavior, and thereby predict it), and *reductionism* (i.e., the notion that we can account for complex natural phenomena in terms of simpler ones). Indeed, they think that Physicalism is *synonymous* with realism, naturalism, rationalism, and reductionism; they believe, to this day, that there is no rational alternative to it. So whatever problems and shortcomings Physicalism may have, it's the only remotely acceptable game in town—or so the story goes.

For someone like you, reading a book that offers just one such a rational alternative to Physicalism, this may come as a surprise. Yet it's true: many scholars are physicalists *merely for a perceived lack of better alternatives*. I know this from personal experience. Mainstream Physicalism has accrued so much cultural momentum that many take it for granted that no rational alternative *can* exist. *Therefore, it's useless to look for one.* These people will thus never bother to pick up a book like the one you're now reading, so the whole thing becomes a self-fulfilling prophecy.

More than once have I debated scholars who came into the debate expecting—no, *knowing*—that Idealism means that the world is supposed to be inside your head. But when it dawns upon them, at some point in the first 20 minutes of the debate or so, that I'm defending a realist, naturalist, rationalist, and reductionist position (I reduce everything to a field of subjectivity, as I'll discuss later), as opposed to a 'spiritual' one, one can palpably see the bewilderment in their eyes. I don't know how abiding the effect is—for all I know, 48 hours later they are merrily back to Physicalism by reflex—but the moment of the realization is clear to any attentive observer; and rather fascinating. If you have watched some of my debates online, you probably know what I am talking about.

The misperception that Physicalism is the only rational metaphysical hypothesis, a self-fulfilling prophecy as it is, helps keep the momentum of Physicalism in both academia and society at large.

A related problem is that many people—scholars in particular—are initially unable to evaluate non-physicalist hypotheses without surreptitiously bringing unexamined physicalist assumptions into the equation. They don't do this maliciously (nobody fools themselves knowingly); they just can't look at things without physicalist glasses on. They are so used to wearing those glasses that they entirely forget they have them on, and thus mistake the tinge of the lenses for established facts.

For example: in a public debate I had with psychologist Susan Blackmore in 2023, she had a visibly hard time reconciling (a) the empirical fact of an external world behaving regularly and predictably, with (b) the notion that this world may be mental. To her, these two things seemed to be mutually exclusive. I believe that, in her mind, anything outside living beings can only be non-mental, for minds can only exist inside biology; and anything mental should behave in whimsical, irregular, and unpredictable fashion, for this is how biological minds operate. So how could the *external, non-biological* world be mental? How could nature, which operates according to regular, self-consistent, and predictable 'laws,' be mental? Even if I am wrong about Susan, I am confident that these kneejerk associations apply to a great many scholars and educated people. And where do these kneejerk associations come from?

From mainstream Physicalism, of course. It is only under Physicalism that minds can only exist inside living beings, for minds are supposed to be somehow generated by metabolism. And since the minds of living beings tend to be rather complex, whimsical, and difficult-to-predict phenomena when compared to inanimate nature, a regular and predictable external world simply cannot be mental—or so the story goes. Now, of course, judging alternatives to Physicalism by presupposing premises that only hold under Physicalism is an instance of question-begging; it's an obvious, textbook logical fallacy.

Highly-evolved living beings—such as higher primates, cetaceans, and pachyderms—have developed complex, higher-level mental functions as a response to environmental challenges. They have these reactive, seemingly unpredictable minds because they evolved within the constraints of a planetary ecosystem that demands *adaptability*. But if inanimate nature is itself mental in essence—i.e., if the 'physical,' inanimate world is merely our internal dashboard representation of a natural mind-at-large—then none of these characteristics should apply to it. For mind-at-large didn't have to evolve in a planetary ecosystem; it didn't have to develop higher-level mental functions such as self-awareness, metacognition, reflection, etc.; it didn't have to adapt to changing environmental challenges. Instead, it stands to reason that it is a comparatively simple and predictable mind; a spontaneous, instinctive, non-reflective mind. And this is why, under Analytic Idealism, nature's behavior is regular and predictable, like that of less-evolved, purely instinctive life forms. The regularity of nature's behavior is thus not at all inconsistent with the idealist hypothesis that nature is mental (on a side note, even human minds are significantly more regular and predictable than non-psychiatrists and non-psychologists may imagine them to be, but I will leave this out of my argument so to acquiesce to vulgar perceptions on the matter). And neither is the notion that mind can exist outside living beings, for, under Analytic Idealism, minds are *not* generated by metabolism; instead, metabolism is simply what some minds *look like* when *represented* on an internal cognitive dashboard.

The problem is that most people will not give themselves the chance to go through even the rather straightforward thought process I clarified above. We live in a society where quick judgment is a survival advantage. Alas, quick judgment, in this case, means begging the question, circular reasoning, cluelessly making unexamined assumptions, etc. And since all these kneejerk assumptions will almost invariably be *physicalist* assumptions, no alternative to Physicalism ever gets a fair hearing. As a result, Physicalism is perpetuated. This is a big thing it has had going for it throughout the 20th century, and to this day.

In popular culture—and even among scholars who should know better—Physicalism also gets credit for the extraordinary success of science and technology since the Enlightenment, even though it deserves just about no credit for either. After all, Physicalism makes statements about what nature *is*—namely, that nature is supposed to have no intrinsic qualities and should be, in principle, exhaustively describable through quantities alone—therefore being a *meta*physics, not a scientific hypothesis. The latter models and predicts nature's *behavior*, being methodologically, at a very fundamental level, agnostic of what nature *is* or *isn't*. All scientific experiments and reasoning can be carried out, in exactly the same way that they have been so far, under any other metaphysics consistent with realism, naturalism, rationalism, and reductionism. That modern science and materialism have more or less concurrent origins in Western culture is entirely—and literally—circumstantial, as Physicalism was simply a convenient political move in early science's fight for survival against the Church. Yet, such concurrence of origins leads many to mistakenly believe that Physicalism underpins modern science methodologically.

When it comes to technology, this vulgar fallacy is even more pernicious. Technologists—i.e., engineers, of which

I am still a practicing one, with more than a few computers, processors, and patents to my name—hardly care about truth in general, let alone metaphysics. We—I'm now speaking with my technologist hat on—operate on the basis of what *works*, not what is ultimately true or not. That's why we use finite-element modeling, Fourier optics, discrete-element transmission line theory, etc., all of which we know aren't true; but they *work in practice*. We just love such convenient fictions that work without over-complicating things. No technological advancement has ever depended on Physicalism or any other metaphysical view. To think they have reflects significant naiveté about how technologists operate and what mindset they have while doing their job.

Nonetheless, the popular belief that Physicalism somehow enables technological development and underpins science is fantastically widespread. People mistake the success of science and technology for the success of Physicalism. So, the more scientific and technological successes we score, the more momentum is imparted on Physicalism, even today. This is yet another psychological dynamic that benefits Physicalism on an ongoing basis.

But the biggest thing Physicalism has going for it today, and counterintuitive as it may sound at first, is probably its *incompleteness* and *vagueness*. In a sane world, these would be reasons to abandon Physicalism; but in ours, incompleteness and vagueness make it impossible for critics to pin down specific physicalist claims and refute them, for the claims that Physicalism does make are almost always blurry, moving, slippery targets.

Notice that, although Physicalism is a metaphysics (*not* a scientific theory), insofar as it maintains that the brain generates the mind it *implies* a scientific hypothesis: namely, that the behavior of the brain causally modulates the behavior of the mind, according to some recognizable mechanism. All scientific criteria then apply: the hypothesis must be formulated explicitly, make testable predictions, and ultimately be falsifiable through experiment. An explicit formulation of the hypothesis will specify *how* brain states modulate mind states. Testable and ultimately falsifiable predictions will specify what reports should be expected from subjects when certain patterns of metabolic activity are observed—or even induced—in their brain.

Specificity is the key requirement here. Although—as we've seen in the previous chapter—Physicalism fundamentally cannot determine *how* brain function supposedly generates experience, it should at least specify what aspects of brain metabolism are *salient* for, or *correlate* with, experience in a consistent fashion, and then explain why. This is the minimum required from any metaphysical theory having such a major neuroscientific implication; anything less than that and what we get is more hand-waving than metaphysical theory.

But Physicalism offers no such specificity. At present, *any* level of brain activity, of *any* kind, unfolding according to *any* pattern, *anywhere* in the brain, measured or reasonably inferred, may be claimed as the physical basis of experience, depending on which data set one is looking at. If *one* neuron *might* be firing *somewhere*, presto, your near-death experience under cardiac arrest has a physicalist explanation. Indeed, physicalist neuroscientists have proposed a cornucopia of different and mutually exclusive causal accounts of consciousness, so to contend with different sets of clinical and experimental evidence. At a very generic level, most do claim that experience vaguely correlates with neuronal activations; but because this is clearly contradicted by significant new evidence, alternative accounts keep on popping up to address the 'anomalies.' The problem is that these accounts are mutually exclusive; they can't all be correct, and thus the totality of the evidence can never be satisfactorily explained under physicalist premises.

For instance, since psychedelic trances are now known to be characterized by *reduced* brain activity—while the richness and intensity of experience go off the scale—neuroscientists working with psychedelics have proposed that experience correlates with functional coupling, or activity variability, or—my favorite for its sheer ludicrousness—minute increases in brain noise, etc., but not with straightforward brain activity. Another group of neuroscientists, working with subjects in regular states of consciousness, equates reportable experience with certain topologies of information integration in the brain, thus also not pure activity. Yet others maintain that experience arises from back-and-forth communication between lower and higher brain areas. And all this is just the start. If you do an Internet search on 'theories of consciousness,' you may be shocked by how many different and mutually exclusive physicalists' hypotheses are out there, being discussed in mainstream academic publications. Hardly any of these hypotheses can ever be definitively falsified—precisely because they are all so vague—and thus there is never closure.

Many of the physicalist accounts of consciousness are not only vague and blurry, but also unstable: they are constantly changing, often from publication to publication, even within a single team. The psychedelic research team at Imperial College London, for instance, in the course of just ten years, has suggested at least three or four completely different theories for how to account for experience in the psychedelic state, all of them mutually exclusive. The same can be said of tentative physicalist accounts of, e.g., memory: they are literally all over the place, some claiming that memory is stored within neuronal structures, others across neuronal networks, in different areas of the brain, operating according to different recall mechanisms, etc. This is great for generating an endless list of topics for publications, PhD programs, funding, vacancies, etc., but it isn't conducive to actual scientific progress, in that it's entirely based on a (false) *metaphysical premise*.

So, if metaphysically-neutral neuroscientists were asked to try to falsify Physicalism on a purely neuroscientific basis—not a philosophical or physical one, for this we have already achieved in Chapter 3—they would first have to ask: *what physicalist hypothesis do we need to test?* Which hypothesis speaks for Physicalism? And even if they could determine which one, chances are that the hypothesis would itself be so fuzzy and ambiguous that no specific tests could be devised to pin it down.

As such, the incompleteness and vagueness of Physicalism render its neuroscientific implications practically *unfalsifiable*. Incredibly, it is precisely by *not* explaining how the brain supposedly generates the mind that the claim that the brain somehow *does* generate the mind achieves a form of experimental immunity. Indeed, the fact that Physicalism *doesn't* explain anything in neuroscience is precisely—surreally enough—one of its key cultural strengths. It keeps the door open to an endless procession of vague and often incoherent hypotheses—commanding plenty of taxpayer money in the process—thus perennially avoiding a defining showdown that could bring resolution.

Another subtle and perhaps unexpected thing Physicalism has going for it, when it comes to its popularity among the general population, is *ignorance*. Yes, ignorance of what Physicalism means, entails, and implies. Many people who casually identify as physicalists—or 'materialists,' the more popular colloquial term—often lack an even basic understanding of what Physicalism means. They tend to think that, according to Physicalism, what is inside their head is just their thoughts, emotions, fantasies, etc.—i.e., their *endogenous* mental states—but *not* the contents of their perception. They think that the colors they see, the flavors and smells they feel, etc., *are really out there in the world*. They think that the world, as it is in itself, is made of the qualities of their perception. But this is not at all what mainstream Physicalism claims.

According to Physicalism, the colors you see, the flavors you taste, the smells and textures you feel, the melodies you hear, *are all inside your skull* because, as qualities of experience, they are supposedly generated by your brain. There is stuff out there, alright, which *corresponds* to these perceptual qualities, but that stuff is not the qualities you perceive; that stuff out there is purely abstract and cannot even be visualized. The best you can do to imagine what the external world is, under Physicalism, is to think of it as disembodied numbers and mathematical equations floating in a vacuum; and even this is way too concrete.

According to Physicalism, the world constituted by the contents of your perception is entirely inside your head. The real inner surface of your skull is beyond the walls you see around you; beyond even the sky you see above your head, insofar as by 'sky' we mean a set of experiential qualities (a blue space with fluffy white clouds, etc.). If you can truly wrap your head around this, you will realize that a vulgar but very popular criticism of Idealism—namely, that under Idealism the world we see around us is inside our head—applies in fact *only to Physicalism*.

But what about Analytic Idealism? Does it follow mainstream Physicalism in claiming that the contents of perception are inside our head? No, absolutely not. Analytic Idealism does say that the contents of perception, for being cognitive representations of the real world outside, are inside our *individual mind*, but not our *head*. This is a huge difference, which may nonetheless require some unpacking.

Under Physicalism, your physical head is a real thing, a real object of the real world. Therefore, it can contain other things.

And so it is that, under Physicalism, the contents of your perception are contained by your head. But under Idealism, the 'physical' world is not the real world; it is instead a *cognitive representation thereof*. Now, your head is part of this 'physical' world. Therefore, it is also a cognitive representation, not a real thing that can contain other things. Your head, as perceived, is an icon on the computer desktop, a dial indication on the dashboard, not a real thing. This way, your head doesn't contain your mental activity; instead, it is *a representation thereof*. This is why, under Analytic Idealism, the contents of perception are *not* inside your head.

Let us again try a metaphor to clarify this. Imagine that we are having a video call online. You are in your home, and I am in mine. But you can see my head on your phone's screen. Well, not quite: you can see *an image of* my head on your screen, a *representation* of my head in the form of colorful pixels. Would you then say that this image of my head contains my thoughts? Of course not; pixels don't contain thoughts; images are mere representations, not containers. The best you could say is that the image in some sense *represents* my thoughts, or is *correlated with* my thoughts in some manner (e.g., through my facial expressions, as represented on your phone's screen).

Now, in exactly the same way, under Analytic Idealism my actual head—the 'physical' entity you could see and touch if you were right in front of me—is but an image, a representation of my individual mind, pixels. In other words, my head is part of what my mind *looks like* when you measure and then represent it on your internal dashboard. As a 'physical' entity, my head doesn't contain my thoughts; instead, it is part of what my thoughts—and the rest of my mental inner life—look like when represented on a dashboard. Do you see the difference? Representations don't contain anything, for the same reason that the pixels representing my head on your phone's screen don't contain my thoughts. Representations just, well, *represent* things.

That my head is made of tiny atoms or elementary subatomic particles means only that such 'physical' representations are, ultimately, also pixelated; just as an image on your phone's screen is pixelated.

Ergo, under Analytic Idealism your head does *not* contain the world of your perceptions; it's the world of your perceptions that contains your head, insofar as your head is a perceived entity. And this is, in fact, exactly what our natural intuition tells us: our heads are in the 'physical' world, not the 'physical' world in our heads. It's so obvious it's almost embarrassing.

Amazingly, thus, casual physicalists attribute to Idealism precisely one of the most counterintuitive aspects *of their own metaphysics* (namely, that the world of perception is inside our head), while attributing to Physicalism one of the most intuitive aspects of Idealism (namely, that the world of perception is *not* inside our head, but our head in the world of perception). It's quite an ironic cultural game of bait-and-switch.

The discussion about Analytic Idealism above is a bit of a digression, in that I am still to elaborate much more carefully on what Analytic Idealism is, what it entails and implies, why you should take it seriously, etc. This will come in later chapters. But I am deliberately dropping hints and partial characterizations of Analytic Idealism as I go along for two reasons: first, to slowly acclimatize you to a different perspective, a different way of thinking about reality; and second, to immediately establish contrasts between mainstream Physicalism and Idealism, so to constantly highlight to you the importance of being critical about metaphysics in general, and Physicalism in particular.

There is one more thing mainstream Physicalism has going for it; one that is particularly pernicious: mainstream media bias, especially the science media. I am not suggesting conspiracies of any form here, and I don't believe there is any (frankly, I don't think the people involved are clever enough to pull off such a thing). What happens is much more banal: good-old human insecurity, laziness, opportunism, careerism, disregard for ethics, and, of course, stupidity.

As I mentioned earlier, research has consistently shown—for over a decade, with results reproduced by many different research groups, using a variety of psychedelic substances and imaging instruments—that psychedelics only *reduce* brain activity, not increasing it anywhere in the brain beyond measurement error margins. I have discussed and documented this *ad nauseam*, in multiple publications, such as my *Scientific American* article with Prof. Edward F. Kelly, titled "Misreporting and Confirmation Bias in Psychedelic Research" (2018), as well as Chapter 27 of my earlier book, *Science Ideated* (2021).

But if you were to read a report published on *CNN* on April 13, 2016—titled "This is your brain on LSD, literally" and authored by journalist James Griffiths—you would have found the following statement: "Images of the brain under a hallucinogenic state showed almost the entire organ *lit up with activity* ... The visual cortex became *much more active* with the rest of the brain" (my emphasis). However, the scientific paper covered in this report—namely, "Neural correlates of the LSD experience revealed by multimodal neuroimaging," by Robin Carhart-Harris *et al.*—stated no such thing. Not only that, it showed *precisely the opposite*: that brain activity *decreases across the brain*, and *across frequency bands*, in the psychedelic state induced by LSD. How can the science media report *precisely the opposite* of what the study has found?

A key illustration in the scientific paper depicts a brain with regions highlighted in yellow, orange, and red. But what those colors mean is an increase in *resting-state functional connectivity* (RSFC) across brain areas, in the psychedelic state. In other words, brain activity is reduced across the brain with LSD, *but the residual activity that is left exhibits relatively more correlation across different brain areas.* The journalist reporting on the study, however, seems to have neglected to read the figure's captions, thereby reporting what he *thought* he was looking at, which happens to have been precisely the *opposite* of what the study found. I have publicly called for a correction of this report multiple times, not only towards *CNN* but also towards the study's authors, who I think have the ethical responsibility to at least try to correct misrepresentations of their work by the media. Nonetheless, as I write these words, almost exactly seven years later, the appallingly incorrect report is still to be found on *CNN*'s website.

But this is just the beginning. Reporting on the exact same study, the *Guardian* newspaper in the UK—in a report titled "LSD's impact on the brain revealed in groundbreaking images," by Ian Sample, science editor of the *Guardian*, on April 11, 2016—reproduced a version of the very figure in question, but with one twist: all references to RSFC—i.e., references that explain what the figure *means*—were removed. Instead, the *Guardian* added the following caption: "A second image shows different sections of the brain, either on placebo, or under the influence of LSD (lots of orange)." As a reader, you *cannot* know what "lots of orange" mean, unless the original figure caption had been reproduced, or replaced with running text. And when you see a brain under LSD depicted with 'lots more orange' than a brain under placebo, you are bound to conclude that LSD lights up the brain like a Christmas tree. Yet the opposite is what was actually found.

The *CNN* report, in fact, reproduces the exact same figure, edited in the exact same way, so as to remove references to RSFC. Both *CNN* and the *Guardian* credit Imperial College London as the source of the image. It is impossible for me to determine which party actually edited the original to remove the references. But if the science institution in question is responsible for the, well, 'simplification' of the figure, I am being unfair to the media and the problem is a lot more alarming than I thought at first. Be that as it may, as of this writing both reports are still online.

Two years earlier, the same group from Imperial College London had published another paper on the neuroscience of the psychedelic state: "Enhanced repertoire of brain dynamical states during the psychedelic experience," by Enzo Tagliazucchi *et al.* (2014). That study found that, in the psychedelic state, brain activity *variability* increases in dream-related areas. In other words, although brain activity *decreases* in the psychedelic state—which many other studies have shown *ad nauseam*—the *residual* activity left behind varies more. The difference between activity and activity *variability* is entirely analogous to that between speed and acceleration: the latter is the first derivative of the former; they are *not* the same thing.

To produce this result, the study did a power spectrum analysis of the brain activity signal read out by brain imaging instruments. Technically, we say that such an analysis happens in the *frequency domain*. And because it does not preserve phase information, one cannot transpose the conclusions to the *time domain*. To put it less technically, the study was such that no statements could be made about the *amplitude* of the brain activity signal in time, which could otherwise have shown how much brain activity there was. All the study could say is whether that activity—however high or low it might be—*varies* more under psychedelics. I am going through the trouble of explicitly naming the technical issues here because I want to impress upon you that, as someone with a background in electronics engineering, I understand the signal processing science behind the study very well; it's rather straightforward stuff for anyone who studied electronic communications.

But several media outlets reported on this study by stating that it found brain *activity increases* in dream-related areas of the brain, in the psychedelic state. For instance, the *Washington Post*—in a report titled "Psychedelic mushrooms put your brain in a 'waking dream,' study finds," by journalist Rachel Feltman, on July 3, 2014—stated: "After injections, the 15 participants were found to have *increased brain function* in areas associated with emotion and memory" (my emphasis). They then went ahead and quoted study coauthor (!) Robin Carhart-Harris in an interview as having said: "You're seeing these [dream-related] areas getting louder, and *more active*" (my emphasis). And: "It's like someone's *turned up the volume* there, in these regions that are considered part of an emotional system in the brain. When you look at a brain during dream sleep, you see the same *hyperactive* emotion centers" (my emphasis). But the scientific paper says nothing of the kind; it says only that activity *variability* increases in dream-related areas of the brain, not activity per se; the methodology of the study makes it structurally impossible to extract conclusions about the latter.

I immediately e-mailed the paper's authors seeking clarification. In private e-mail correspondence, first-author Enzo Tagliazucchi confirmed to me, in writing, that I was correct. Involved in the correspondence, Carhart-Harris acknowledged that he indeed had misunderstood the conclusions of the study. I find this plausible, as Carhart-Harris doesn't seem to have a background in signal processing, and the technical issues involved can be tricky for a layperson. Nonetheless, I fully expected that he and Tagliazucchi would promptly issue public corrections, since the incorrect statements were prominently featured by the media and scientific integrity demanded decisive action. But no such correction has ever been made, despite my publicly calling for it repeatedly over the years.

Not only that, Carhart-Harris penned a popular science essay on the science blog *The Conversation*, where he insists on the error: "psilocybin increased the *amplitude* (or '*volume*') of activity in regions of the brain that are reliably activated during dream sleep" (my emphasis), he wrote. This is flat-out and unreservedly false: amplitude can only be determined in time-domain analyses, not from a power spectrum lacking phase information (technically, without phase information one cannot know whether the frequency components interfere constructively or destructively with one another, in the time domain). As I write these words, in 2023, one could still find this quote in his essay "Magic mushrooms expand your mind and amplify your brain's dreaming areas – here's how," published on July 3, 2014.

He has argued by e-mail that the word 'activity' is ambiguous and, therefore, could be interpreted as activity *variability*. I believe this is brazenly false and insisting on it is, at the very least, irresponsibly and gratuitously misleading; for any educated reader will read 'activity' as, well, *activity*. Moreover, later in the very same essay, Carhart-Harris does use the word 'activity' in the correct, normal sense: "Our first study, published in Proceedings of the National Academy of Sciences in 2012, revealed decreases in brain *activity* after injection of psilocybin" (my emphasis). What the paper cited shows is a proper reduction in *activity*, not activity variability. It therefore continues to baffle me that, almost a decade later, none of these errors—including those in an essay penned by Carhart-Harris himself—have been corrected.

It gets worse. A few years later, in a social media exchange in which I, once again, publicly called on the authors to issue corrections, Enzo Tagliazucchi stated that it was *me* who misunderstood the implications of their signal analysis, mistaking frequency-domain for time-domain conclusions. He wrote on X, then known as Twitter: "We had a long time ago the same discussion when you confused changes in BOLD activity levels with changes in BOLD variance" (this tweet, published at 6:26 PM CET on October 27, 2018, is still online as of this writing). Astonishingly, Tagliazucchi was accusing *me* of the very error his coauthor Carhart-Harris had made towards the media, which *I* had pointed out to them four years earlier! This, of course, prompted me to immediately publish the full record of the correspondence we had had on the topic, on my personal blog, under the title "Setting the record straight with Robin Carhart-Harris and Enzo Tagliazucchi," on October 28, 2018.

I do not believe that malice was involved in any of the original errors (though I reserve judgment about the continuing unwillingness of the people in question to publicly correct those mistakes). I believe that Carhart-Harris honestly misunderstood the signal analysis that his own colleague had performed and reported. I also believe that Tagliazucchi sincerely misremembered who was confused about what. Yet, *how* could he not only misremember something so simple, but *invert* the facts? *How* could Carhart-Harris communicate towards the mainstream media, with obvious confidence and authoritativeness, something he had not understood at all—because he lacks the background—and should have known better? Here's your answer: since Physicalism *has to be right*—we all know it is right, don't we?—it *must* have been the case that brain activity increased in dream-related areas; for this is *exactly* what physicalists would have expected to see, due to the similarities between psychedelic and dream states. The fact that the study's methodology *cannot*—and was never meant to—measure the amplitude of brain activity played a minor *de facto* role in face of such powerful theoretical prejudices. Moreover, since non-physicalists are always wrong—we all know that too, don't we?—it must have been Bernardo Kastrup who got confused about the methodology four years prior, not a physicalist coauthor of the very study in question.

In science we call this 'confirmation bias,' this case being, in my opinion, a spectacular and grotesquely sustained instance thereof. This is why I considered it worthy of recounting in detail: it makes the otherwise abstract notion of confirmation bias very concrete; it brings it to life in all its rawness. Had I not told you about these specific, real-life facts, I might have left you with the impression that my claims about confirmation bias are merely vague, generic, unsubstantiated, and perhaps even biased themselves. But with concrete, living *examples*, such as the above, I hopefully could impress upon you the alarming reality of confirmation bias.

More generally, confirmation bias means this: when you expect certain results or conclusions with deep but unexamined and uncritical confidence, you will see—and therefore report— what you expect to see, regardless of what is actually in front of your eyes. You will also design your experiments to find what you expect to find, as opposed to what is there to be found. Finally, you will do your statistical analysis so as to carve out and highlight what you expect to be the case, and filter out more significant effects that you consider *a priori* impossible. You may even read into your results—and thereafter communicate to the press—what you believe they should imply, instead of what they actually imply. And once you are publicly caught in this web, it becomes difficult to disentangle yourself from it without harm to your reputation and career; the only perceived option is to double-down. Therefore, errors are perpetuated; no one corrects anything; one just hides behind precarious and lamentable language games, such as the notion that the word 'activity' is so ambiguous as to mean variability, or functional connectivity, or entropy, or whatever else you find increasing in the brain under psychedelics. Under the pretext of making things 'simpler and easier for intellectually stunted science journalists,' crucial details are omitted in press materials, which allow for a result that clearly *contradicts* physicalist expectations to be misinterpreted—and then misreported—as *corroborating* these expectations. This is the world we live in. The confirmation bias injected into science by the metaphysics of Physicalism is probably the most powerful there has ever been.

But I don't want to leave you with the impression that the problem is localized and contained. So I shall give you one more example. In a 2014 scientific paper titled "Homological scaffolds of brain functional networks," by G. Petri *et al.*, researchers tried to illustrate how the correlations across residual activity in different brain areas increase under psychedelics (even though activity itself decreases). To do this, they used graphs with linked nodes, different nodes representing different brain areas, and the links representing the correlations between the respective areas. They then applied *successively lower* thresholds of correlation for drawing the links, until enough links—whatever 'enough' means in this case—were visible in the graph. This, of course, created the arguably artificial and misleading appearance that the brain under psychedelics exhibits dramatically increased global connectivity.

In all fairness to the researchers, in the scientific paper they warn that the graphs are but "simplified cartoons," encouraging caution in their interpretation. However, these same graphs were subsequently used, *with no such qualification*, by respected journalist Michael Pollan in his book *How to Change Your Mind* (Penguin, 2018) as the primary 'evidence' for a physicalist interpretation of the results. Puzzlingly, Pollan barely mentions the far more impressive and direct measurements of decreased brain activity reported in multiple other studies. In how many other popular science books, newspaper reports, and magazine essays, do you think confirmation biases such as this can be found? And how many of them pass for perfectly good and trustworthy science material?

The problem here is not theoretical or abstract; it's becoming increasingly more concrete and alarming as we transition from Internet search engines to AI queries, such as OpenAI's ChatGPT chatbot. Indeed, traditionally, when using the Internet to find an answer to some question, we did so through search engines like Google. We knew that many search results were questionable, unreliable, or even flat-out malicious, and thus proceeded with caution. We *took responsibility* for finding the correct answer, for we knew that doing so depended on the quality of our search and our ability to critically evaluate the results. We didn't expect that Google would always give us truth; we knew it couldn't do that.

But as we begin to abandon search engines and pose, instead, questions to AI chatbots in natural language, our psychology leads us to believe the answers produced by the chatbot as if they were truths pronounced by an all-knowing oracle who understands what it is saying. The reason is that we no longer see search engine results; we no longer see the *sources* of the information being conveyed to us. Instead, we see a human-like answer on the screen, as if an authoritative professor were verbally elucidating the issue. And since often that answer is true, we slowly let our guard down and drift towards trusting *all* answers uncritically. This is similar to how we slowly drift towards trusting autonomous driving software, even if in the beginning we feel rather edgy about letting go of the steering wheel. The result is that we lose our ability to be critical of the answers and begin to believe in fallacious and even malicious material.

Linking this back to our discussion on media bias: AI chatbots collect the information they provide you with by crawling the Internet. They will look at magazine articles, blogs, news reports, etc., so to find the information you are looking for. And when those sources are biased and wrong, so will the answers the chatbots give you. For chatbots, despite being considered instances of 'artificial intelligence,' in fact have no understanding of what they are saying; none whatsoever. They are merely natural language interfaces to search engines. As such, even if their algorithms favor sources with high credibility scores—such as *CNN*, the *Guardian*, *The Wall Street Journal*, and the academic blog *The Conversation*, all of which are much more credible than your neighbor's social media feed—the nonsense will still make its way to you, because of reporting biases at the highest echelons of academia and journalism. It is crucial for our future as a society that we understand this.

To test my own point above, I decided to ask ChatGPT a question (in May 2023): "Do psychedelics *increase* brain activity?" ChatGPT's answer: "Yes, psychedelics can increase brain activity ... [they] can lead to increased activity in ... *the default mode network* (DMN) ... It should be noted that while psychedelics can increase brain activity in certain regions, they can also lead to decreased activity in other regions. For example, research has shown that psychedelics can decrease activity in *the default mode network*" (my emphasis). Obviously, this answer cannot be correct, as it is internally contradictory; and it is so in a very specific—not a merely generic—manner. Clearly, ChatGPT has no understanding of what it is saying; it simply provides a natural language interface to search results. In reality, research has consistently shown that psychedelics reduce brain activity, *primarily and precisely in the Default Mode Network*. Where do you think ChatGPT's 'confusion' comes from?

AI chatbots are not intelligent in any way remotely akin to how you and I are intelligent. AI chatbots do not *understand* anything; they just collect and present information in natural language format; they reword back to us what is written out there in the wilds of the Internet. So when the media reports on scientific results in an inaccurate, biased manner—or worse, when the researchers themselves do so—those errors and biases are incorporated into our cultural database; the 'oracle' that the vast majority of the population will be consulting to inform their lives for the next decades. This is perhaps the biggest thing Physicalism will have going for it in the future.

But there is more. The psychedelic science cases of media and confirmation bias discussed above are particularly appalling, but countless seemingly innocent instances of the same bias are happening literally every day. Let us take memory, for instance: a core premise of Physicalism is that memory is information physically stored somewhere in the brain, just as your files are physically stored in your computer's main storage drive. This premise has a scientific implication: we should be able to *find* memory information in physical brain states.

Science has been trying to find this secret information storage for well over a century, with results that often contradict the physicalist premise. For instance, in 2013 researchers reported on an amazing study: little aquatic flatworms called 'planaria' — which have the remarkable ability to regrow amputated body parts, including their head — were trained to navigate an irregular surface to find food. The researchers then decapitated the planaria — thereby removing their neurons, which are in their head — and waited for two weeks until a new head grew. With a brand-new head in place, the planaria maintained their originally trained ability to navigate the rough surfaces to find their food, without additional training. Somehow the planaria remembered their training even after their head was severed, contradicting the premise that memories are physically stored in (networks of) neurons. After all, if you throw your computer's main storage drive away, you will not expect that the brand-new one you just bought will automatically have all of your old files in it. The research in question has been published in the paper "An automated training paradigm reveals long-term memory in planarians and its persistence through head regeneration," by Tal Shomrat and Michael Levin, published in the *Journal of Experimental Biology* in October 2013.

But I digress. The point I am trying to make is not *per se* that Physicalism is wrong when it comes to memories — though it clearly is — but that scientific results about memories are reported inaccurately and with bias in favor of Physicalism. To see this, let us take a press release put out on September 10, 2008, by UCLA Health: "How memories are made, and recalled."

The press release ambitiously claims that "scientists at UCLA and the Weizmann Institute of Science in Israel have recorded individual brain cells in the act of calling up a memory, *thus revealing where in the brain a specific memory is stored and how the brain is able to recreate it*" (my emphasis). But when one actually reads the technical paper, a very different picture emerges.

Here's what the researchers did: having instrumented subjects with electrodes to record neuronal activity, they asked the subjects to watch short video clips. They then recorded the patterns of neuronal firings during the experience of watching the video clips. Some time later, the subjects were asked to remember what they had watched. Their patterns of brain activation were recorded again. Lo and behold, roughly the same patterns of neuronal firings were observed during the primary experience *and* the recall of the experience.

Is this result at all surprising? We have known for a long time that many types of subjective experience correlate with specific patterns of brain activation (except, of course, psychedelic states, syncope, hyperventilation states, brain damage associated with acquired savant syndrome, states of cardiac arrest, etc.; but I digress again). Therefore, insofar as the experiences of watching a video clip and recalling the video clip are qualitatively similar, of course the associated patterns of neuronal activation should be similar; duh. This says precisely *nothing* about the supposed physical mechanisms underlying memory.

The relevant question here is this: how does the brain know *which neurons to reactivate* during the recall experience? How does it remember which neurons were active during the original experience of watching the video? Where is the information stored, and later retrieved, which tells the brain which neurons to reactivate during recall? That's what would say something about where memories are stored in the brain. But the study says nothing about it. That the same neurons go active during recall is at least largely irrelevant to elucidating *how* the brain knows which ones to reactivate. The press release's claim—a *UCLA* press release, mind you—that researchers revealed where in the brain a specific memory is stored is unjustified. Indeed, it takes some very charitable imagination jiu-jitsu to figure out a sense in which the claim can relate to the actual experiment.

But we live in an age where doing science is a career job, and research institutions fight tooth-and-nail for funding. Scientists go through performance appraisals every year. Funding and careers can only be secured if visible and relevant progress is made and announced to the world with great fanfare. So people and institutions have every motivation to exaggerate and mislead when communicating to society at large, to highlight the great progress and relevance of their results; without it, there may be less funding and less jobs next year. And since the foundational metaphysical assumption of our culture is Physicalism, the exaggerations and biases consistently toe the line of Physicalism; otherwise, they would attract the wrong kind of attention and scrutiny. Nobody wants that.

Still on the topic of memory, another research group announced, in 2009, the discovery that memory storage may be associated with the interplay between synaptic activity and DNA transcription in the nucleus of neurons (see "Reducing memory to a molecule: A researcher explores the molecular essence of memory," by M. Hendricks, *Johns Hopkins Institute for Basic Biomedical Sciences*). A 2012 study, on the other hand, discovered that memories may be encoded digitally in neuronal microtubules—structures in the neuronal cytoplasm, not in the nucleus (see "Cytoskeletal signaling: Is memory encoded in microtubule lattices by CaMKII phosphorylation?" by T.J. Craddock *et al.*, published in *PLoS Computational Biology*). Yet another 2012 study announced the discovery that memories may be stored as patterns of inter-neuron synaptic connections in the hippocampus (see "Synaptic conditions for auto-associative memory storage and pattern completion," by E.Y. Cheu *et al.*, in the *Journal of Computational Neuroscience*).

All these studies claimed fundamental scientific advancements confirming the physicalist premise that memories are stored physically in the brain. The only problem is: the conclusions are mutually contradictory.

Don't get me wrong: there *are* true and relevant scientific discoveries in probably most of these studies. The problem is that, to highlight their relevance, broad metaphysical claims are made that are often entirely unsubstantiated by the results. For instance, researchers often fail to distinguish memory *pathways* — i.e., neuronal mechanisms correlated with memory *access* — from memory *storage*. If a person becomes unable to recall short-term memories because of damage to certain areas of the brain, maybe 'physical' structures correlated with memory *access* have been compromised, not memory storage itself (otherwise, the well-known phenomenon of 'terminal lucidity' — look it up — wouldn't be possible). But alas, this kind of careful and rigorous reasoning is all but absent when it comes to the science media and the press offices of research institutions. The result is a circus of misleading, biased, and sometimes flat-out false claims being pumped out to and by the media, in an effort to successfully compete in the funding — and career — marketplace.

If you are a casual educated reader simply going over the latest science headlines, you are bound to come away convinced that the physicalist premise that memories are stored physically in the brain has been scientifically proven over and over again. Here are a couple of the headlines published in association with the studies mentioned above: "Reducing memory to a molecule," "Scientists claim brain memory code cracked," and so on. So many times will you have read similar headlines, claiming major advances in pinning down the physical location and mechanisms of recall, you just won't think it possible that they are all wrong. Yet, as far as metaphysics is concerned, they more than likely are, not least due to the fact that the claims are mutually contradictory.

The issue is that Physicalism, for being mainstream, provides a cover of protection for journalists and scientists alike: if you interpret and report results along physicalist lines, you are less likely — or so the calculation goes — to go wrong and be punished for it than if you dare to contradict the mainstream narrative. The latter will attract critical scrutiny, put you on the defensive, motivate scorn by your colleagues (who are competing against you for that promotion or that funding, and will smell blood the moment you hint at a non-physicalist position), entice the public ire of big-mouthed pundits who do no science but make a career of talking about it, dramatically reduce your chances of getting a paper or report published, and so on. At the very least, it will triple the amount of work you will need to do to defend your point. In a newsroom, the difference between safely cowering under physicalist cover or critically questioning it is that of having your report published immediately, or seeing it escalated to management, scrutinized meeting after meeting, having more work demanded of you, and all this for a much higher risk of rejection.

If you go with Physicalism and err, you will be forgiven, for how could *anyone* have imagined that things aren't quite like Physicalism predicts, right? But if you go against it and err, God help you; you will be branded a nut with untrustworthy judgment and face an immediate career glass-ceiling. Therefore, if you have big career ambitions and less big commitments to ethics, you may calculate that it's best to confirm general expectations, even when you know that you are misleading the public and misrepresenting your own results. People willing to do so are the ones who stand the greatest chance of finding themselves in influential and well-paid positions — department heads, chair professors, and whatnot — for obvious reasons. Universities and news organizations compete in an arena where schmoozing with power is king, and power is older; power is old-fashioned; power is *physicalist*. Ergo, firing a shot against it is akin to shooting yourself in the foot. It takes great intellectual, scientific, and journalistic integrity to objectively pursue reason and evidence in a rigorous manner, whatever the cost. Alas, such level of integrity appears to be rare. The more popular calculation seems to be that it's best to err on the side of Physicalism. This is what perpetuates our current metaphysical insanity like a self-fulfilling prophecy, to the point that—scandalously—the *very opposite* of the true findings is what sometimes gets reported.

Clearly, thus, despite being the worst metaphysics on the table today—the most internally inconsistent to the point of incoherence, the most empirically inadequate, having the weakest explanatory power, etc.—culturally speaking Physicalism has *a lot* going for it: a perceived (and self-fulfilling) lack of rational alternatives, the prevalence of unexamined physicalist assumptions leading to widespread question-begging, the delusion that Physicalism underpins science and technology methodologically, the vagueness and resulting unfalsifiability of Physicalism, general public ignorance of what is entailed and implied by Physicalism, and brazen confirmation bias in science and the media in favor of Physicalism. This is why it continues to dominate our culture, despite its untenability being so self-evident and overwhelming to those who care to look a little closer. But who will have the time and background to look closer, unless they are a philosopher specialized in the subject? Who will be able to muster the required energy and discipline, in late evening hours after arriving home from a tiring day at work, to critically study the relevant, often highly technical literature? And thus the colorful carnival wheel of metaphysical drivel, ironically cladded in the garments of science and reason, continues to go merrily round.

Chapter 5

The remedy is worse than the disease

In his *Lectures on the Philosophy of History* (1837), Georg Hegel introduced into the Western mind the idea of historical *evolution*, the notion that humanity somehow progresses in the course of time, advances, gets better in some significant sense, steadily minimizing some error or cost function. Charles Darwin latched onto the underlying intuition in his own *On the Origin of the Species* (1859) — indeed, as Friedrich Nietzsche stated in *The Joyful Wisdom* (1882), "without Hegel, no Darwin" — and cemented steady evolution as one of the most powerful, fertile, and consequential ideas in Western thought. Early 20th-century Positivism, Marxism, Scientism, New Thought, etc., all have their inception in this now deeply ingrained intuition of linear, monotonic increases in human insight.

Never mind that Thomas Kuhn already comprehensively demolished such a naïve and arbitrary idea in his seminal book, *The Structure of Scientific Revolutions* (1962), and that Carl Jung already provided a more empirically adequate model of the dynamics of our epistemology — namely, 'circumambulation' — than linear evolution: today, we still like to regard all mainstream scientific and philosophical developments since the Enlightenment as linear steps forward. Even if earlier Enlightened models or ideas turn out to be false, we find ways to regard them as merely *incomplete* steps that, nonetheless, played a positive role in opening the doors to a deeper understanding.

Allow me to belabor this point a bit, for it is crucial to our understanding of ourselves, and of how we go about the business of metaphysics. We like to see our errors as things we can *build upon*; we like to conjure up and attribute value to all our endeavors, even when they turn out to be just silly. Our psychology has made it extremely challenging for us to recognize that, sometimes, we are just flat-out wrong and that's all there is to it. It makes us uncomfortable to think that certain ideas we've adopted as a culture were just a royal waste of time that led to nothing. We feel the need to validate even our worst missteps, because in doing so we validate *ourselves*; we get that warm and fuzzy feeling that our efforts somehow always lead to *something* of value, incomplete as they may be.

As the obvious shortcomings of mainstream Physicalism begin to erode the confidence of even the most prejudiced and pigheaded intellectual elites, the need to regard it as *incomplete*—as opposed to just silly—is now reaching overwhelming levels. And so it is that, since the late 1980s or so, an effort has been underway in academia to reframe Physicalism as a merely *insufficient* step in the right direction. This effort goes by the name of 'Micro-Constitutive Panpsychism'—or simply *Panpsychism*, as I shall call it henceforth.

There are two different formulations of Panpsychism. I'll first discuss their differences and then elaborate on what they have in common.

The first formulation is the notion that elementary subatomic particles (henceforth simply 'fundamental particles'), *in addition to* physical properties such as mass, charge and momentum, also have *fundamental experiential properties*. In this view, experience is simply *added* as an extra property of matter, next to the other known ones.

The second formulation of Panpsychism, on the other hand, states that experience is the *intrinsic nature* of fundamental particles; that is, what we call a fundamental particle essentially *is* experience, which then manifests itself outwardly, through interaction with other fundamental particles, in the form of the other known properties, such as mass, charge, momentum, etc.

Notice that these two formulations really are different. In the first case experience is simply another property *of* matter, while in the second case matter *is* experience. Yet, their underlying spirit is the same. Indeed, what makes both formulations panpsychist—as opposed to idealist—is this: in both cases, *the structure of experience*—i.e., of subjectivity itself—*is supposed to be the structure of the distribution of fundamental particles across space and time.* In both cases, the world is constituted by these fundamental particles, which, in turn, are the carriers and subjects of experience. Therefore, the way the particles arrange themselves in space and time *is* the way subjectivity arranges itself.

In addition, under both formulations of Panpsychism there is something it feels like to *be* a fundamental particle; there is something it feels like to *be* an electron, a quark, or a photon: electrons, quarks and photons not only have standalone reality, they also are *conscious* in and of themselves; each photon, each quark, each electron is a little microscopic subject of experience, or 'micro-subject.' Our conscious inner life is supposed to be the aggregate result of how these discrete micro-subjects arrange themselves and interact with one another inside our body. The micro-subjects constituting our brain somehow *combine* with one another inside our skull to give rise to our seemingly unitary subjectivity—i.e., to what it feels like to be us.

The motivation for Panpsychism is obvious: since mainstream Physicalism has predictably failed to explain (qualitative) experience in terms of (purely quantitative) physical stuff, the panpsychist just deems the existence of experience to be a brute aspect of physicality that requires no explanation; and presto, no more 'hard problem of consciousness'! The idea is to stick to the same irreducible, basic building blocks of nature that mainstream Physicalism already postulates. In the mind of the panpsychist, these basic building blocks are the fundamental particles: microscopic little 'marbles' that have equally irreducible properties—i.e., properties that cannot be explained in terms of anything else. The panpsychist then simply postulates that experiential states are either (a) just one more type of fundamental property of the particles, essentially extrapolating the line already drawn under Physicalism; or (b) the inner essence of the particles, which fills in a hole left open by Physicalism.

The critique of Panpsychism that follows presupposes only what the two formulations discussed above have in common. Therefore, I shall henceforth no longer differentiate between them.

Notice that Panpsychism explains *precisely nothing*: it simply finds a subterfuge to avoid the need for an explanation. If this were to be considered a valid line of reasoning, then *any* vexing empirical observation could be trivially 'accounted for' in the same manner: How do we make sense of the fine-tuning of the universal constants? Well, it is just a brute fact of physicality. How can we account for why people with no risk factors still develop cancer? Well, it is just a brute fact of physicality. And so on. Is this in the spirit of philosophical and scientific inquiry? (If you think Analytic Idealism—for also considering experience fundamental in nature—is guilty of the same sin, then read on; I will tackle this question very explicitly later. For now, if you are so inclined, you could try and think about it yourself, so we could compare notes later.)

The panpsychist acquiesces uncritically to our psychological need to regard our errors as useful steps on a hallucinated, dearly longed-for, heroic ladder of linear epistemic evolution: it insists that (a) irreducible physical entities—just as under Physicalism—do have standalone existence, despite overwhelming laboratory evidence to the contrary (as discussed earlier); and that (b) the structure of nature, and of subjectivity itself, is indeed—again as postulated by Physicalism—the structure of the 'physical' stuff that appears on the screen of perception. In other words, the panpsychist buys wholesale into the illogical and untenable physicalist notion that the structure of what is represented is the structure of the representations; that, because my image on the screen of your phone during a video call is pixelated, then I must be made of little rectangles myself. And since the screen of perception is just our internal dashboard of dials, the panpsychist—just like the physicalist—continues to maintain that nature out there is dashboard-like; that the clouds, storms and winds outside the aircraft have the shapes of the dials and indicators on the aircraft's dashboard.

Under a panpsychist optic, Physicalism was a necessary step on the road; it got the important things right but just failed to take the last step: to understand experience as fundamentally associated with particles—the fundamental building blocks of nature, in the mind of the panpsychist—as opposed to an epiphenomenal effect of their arrangements and dynamisms. There is a sense in which the naiveté of such a perspective is heartwarming and endearing, like children trying to explain away their misdeeds when they are caught red-handed.

Another philosophical shortcoming of Panpsychism is that there is no explicit, coherent account—not even in principle— of how two or more fundamentally distinct micro-subjects could combine to form a higher-level one. How could the little subjectivities of the myriad particles constituting your brain combine to give rise to *you*, as a seemingly unitary conscious subject? After all, your neurons don't even touch one another, their communication taking place through neurotransmitter molecules that drift across the gaps between them. As a matter of fact, a compelling case has been made by philosopher Sam Coleman that the combination of otherwise separate micro-subjects is an incoherent hypothesis already in principle; as much a hand-waving appeal to magic as attempts to solve the 'hard problem' under physicalist premises (see "The Real Combination Problem: Panpsychism, Micro-Subjects, and Emergence," published in *Erkenntnis*, 2014). If this is true— and I believe it is—then Panpsychism represents no advance in explanatory power when compared to Physicalism; it's the same dead end, just in a different guise.

The philosophical shortcomings of Panpsychism, however, are rendered entirely redundant by a simple scientific fact: *Panpsychism contradicts known physics and is, therefore, demonstrably false.* Indeed, the foundational premise of Panpsychism is that fundamental particles are irreducible entities with discrete spatial boundaries, like little marbles localized in space. This is supposedly the reason why the little marbles inside your head combine to form *your* consciousness, while the little marbles inside my head, in a different spatial location, combine to form *my* consciousness, separate from yours. The spatial boundaries of our respective marbles render your experiential field disjoint from mine, thereby preventing us from accessing the contents of each other's minds—or so the story goes.

The panpsychist cheerfully but cluelessly falls for the most easily avoidable epistemic trap in the book: to run with words stemming from another field—in this case, the notion of 'particles' stemming from physics—without understanding what these words mean and how they are used technically. Indeed, Panpsychism wouldn't even have started if panpsychists grasped what a 'particle' means in modern foundations of physics. The notion that little marbles are the building blocks of nature may have been popular among some pre-Socratic philosophers—such as Democritus—two and a half thousand years ago, but since then we've learned a thing or two. The panpsychist is a little outdated in their grasp of physics.

Indeed, we've known at least since the late 1940s (arguably even as early as the late 1920s), with the advent of Quantum Electrodynamics, that what we call 'particles' aren't particles at all: they are merely local patterns of excitation of a spatially unbound quantum field. Think of particles as ripples on a river: each ripple has a certain height, thickness, speed, and direction of movement, which are the ripple's 'physical' properties. They also have defined locations in space: you can point at a part of the river and say, "There's a ripple!" Yet, *there is nothing to the ripple but the river itself.* The ripple is not a standalone entity, but a behavior of the river; it's not a thing but a doing. This is why you cannot grab a ripple and lift it off the river.

In precisely the same way, Quantum Field Theory (QFT)— the more general formulation of Quantum Electrodynamics, which also happens to be the most accurate and exhaustively confirmed scientific theory ever devised—tells us that the so-called particles are just 'ripples' of a quantum field. In essence, *there are no true particles;* we use this word today only metaphorically, as shorthand, and for historical reasons. There are only quantum fields, which are spatially unbound. 'Fundamental' or 'elementary' particles are 'fundamental' and 'elementary' just in that they are not constituted by *other particles* (in the way that, e.g., protons and neutrons are constituted by quarks and gluons, only the latter two being 'fundamental' or 'elementary'); not that they are irreducible. For almost a century, physics has known that particles and their properties are reducible to fields.

Therefore, if the panpsychist wants to avoid the 'hard problem' by making experience, subjectivity, consciousness itself, a fundamental property or essence of an irreducible physical entity, then only a *field* can be that entity; for the field is all there is. It is the *field* that must be conscious, not a particle, for there's nothing to the particle but its associated field; just as there's nothing to a ripple but the river that ripples. Taken in by the most superficial of semantic mix-ups—by a mere word—the panpsychist sees *things* where there are only *doings.*

Once this semantic error is cleared up, Panpsychism implodes. After all, the same spatially unbound quantum fields span the space occupied by your body and mine. So why can't I read your thoughts and you mine? How can our respective *experiential* fields be disjoint, if the same *quantum* fields underly—as they do—you and me? Panpsychism collapses the moment it is rendered in a physically consistent manner.

Some panpsychists, such as philosopher Philip Goff, seek refuge from this inevitability in Bohmian Mechanics, a niche interpretation of quantum mechanics that preserves the marble-like nature of particles. But this only reflects continued lack of familiarity with the basics of modern physics. Even if Bohmian Mechanics hadn't been experimentally refuted a few years ago (see a summary of these results by Natalie Wolchover, in a piece titled "Famous Experiment Dooms Alternative to Quantum Weirdness," published in *Quanta Magazine* on October 11, 2018), it does not have a relativistic extension to reconcile it with Special Relativity. This alone renders it untenable, for Relativity has been experimentally confirmed *ad nauseam*. Indeed, it can be argued that *any* formulation of Quantum Mechanics that *can* be reconciled with Relativity will, *perforce*, entail a field-excitation understanding of particles.

It is also this field-excitation understanding that allows QFT to make sense of a great many empirically observed phenomena, such as:

(a) The spontaneous appearance and disappearance of particles in a vacuum—the so-called 'quantum fluctuations'—which would amount to magic if particles truly were little marbles: how could these marbles magically cease to exist, and then again pop into existence out of nothing? But if particles are simply field excitations, then the magic disappears: nothing pops in or out of existence; what happens is that an underlying field, which is always present, sometimes 'ripples' (i.e., becomes excited) and sometimes goes back to rest (i.e., becomes unexcited). Rivers, too, sometimes ripple and sometimes don't.

(b) Particle *interactions*, which are left out of classical quantum theory but have been very successfully modelled under the premises of QFT: 'ripples' of underlying fields sometimes collide and interfere with one another in predictable ways, leading to the formation, disappearance, or modulation of other 'ripples.' That's what the famous 'Feynman Diagrams'—which earned physicist Richard Feynman a Nobel Prize—so successfully show. Without a field-understanding of particles, we cannot account for how they interact with one another; i.e., we cannot account for essentially *anything* that happens in nature.

(c) Spontaneous particle decay—i.e., the fact that a particle can spontaneously become other particles that weren't part of the original one to begin with. For instance, a Higgs boson can decay into two muons. But the Higgs boson is not made of two muons, so where did the muons come from and, just as importantly, where did the original boson go? Well, a Higgs boson is just a 'ripple' on the underlying Higgs *field*. That 'ripple' has specific properties—things analogous to the height, breadth, speed, and direction of movement of a ripple traveling on the surface of a river—that define it as a Higgs boson. But with time, that original ripple loses energy, becoming e.g., shorter, broader, traveling slower, etc. It may even turn into two or more other ripples if it encounters an obstacle or transits from the river into a broader lake through a narrow channel. These new ripples with different properties are different particles, and that is what particle decay is; that's why the Higgs boson can decay into two muons: both the boson and the muons are just ripples with different properties, not marbles that magically disappear and reappear in different forms. (Incidentally, contrary to popular belief, we have never measured a Higgs boson directly; it is too unstable and decays before it can interact with a measurement surface. What we *did* measure at CERN were the particles the Higgs boson decays into, which then allowed us to retroactively infer the existence of the boson.

Particle decay, which cannot be made sense of without a field-excitation understanding of particles, is entwined with the entirety of modern high-energy physics. To give up on this understanding is thus to give up on all high-energy physics, the foundation of all sciences.)

(d) Everyday regularities of nature that can be observed without instrumentation, such as the defining behavior of mass called 'inertia.' Indeed, inertia—the resistance of mass to changes in the speed and direction of its movement—can only be made sense of under the understanding that the Higgs boson betrays the existence of an underlying field: inertia is the effect of mass trying to go through the 'viscous' Higgs *field* in which it is always immersed, like fish in water. Without this field—i.e., without the viscosity of the 'water'— all the 'fish' would be swimming at the speed of light and there would be no passage of time. The relevance of the discovery of the Higgs boson for our understanding of the origin of mass—which is what motivated a gullible press to call the boson the 'God particle'—thus rests on the fact that it shows a 'ripple' on the Higgs field; it is the *field* that explains the existence of mass.

(e) Etc.

If we were to give up on the understanding that particles aren't marbles but local field excitations instead, all the above phenomena would suddenly become magical. Methinks we aren't going to do that for the sake of the panpsychist. Therefore, it shouldn't surprise anyone—except perhaps the panpsychist— that Bohmian Mechanics was quickly renounced *by its very creator*—Louis de Broglie—already a century ago.

Because of all the above, appeals to Bohmian Mechanics just make the panpsychist look silly. Yet, without such appeals, Panpsychism remains physically incoherent and, therefore, a nonstarter.

Having beaten this dead horse for long enough, we can now turn our attention to a more interesting and productive question, as it can help us gain insight into the thinking processes behind our mistakes: what are the *intuitions* that render Panpsychism so appealing to some and what, *precisely*, is wrong about them?

The core of the panpsychist's intuition is that we, subjects of experience, appear to be *compound entities*. In other words, we are seemingly made of proper parts, such as discrete living cells, put together to form our body and brain. After all, every person is a subject, and every person has a body made of cells. Moreover, these cells are themselves compound entities as well, in that they are constituted of numerous fundamental particles put together. As such, our very consciousness—or so the intuition goes—must also be compound, somehow arising from the combination of lower-level constituents. This *does* sound logical at first, doesn't it? The problem, of course, is that proper philosophy calls for more than just superficial appearances.

More than one unexamined assumption, mistakenly taken for fact, underly the panpsychist's intuition that consciousness—subjectivity itself—must be compound. For starters, that the body is a compound structure does not entail or imply that the subjectivity associated with the body is itself compound. We already briefly touched on this point above but, in the spirit of Jungian circumambulation, let's revisit it in more depth: a biological body is a perceptual *representation*, a thing we see, feel, smell, etc. But the structure of representations on the screen of perception is not necessarily the structure of the *subject of perception* (why would it?). Allow me to repeat this for clarity: the panpsychist mistakes the structure of the *contents of perception* for the structure of the perceive*r*. Conflating these two things leads to category mistakes.

To see why, let's revisit an analogy we've discussed earlier: if I were to talk to you remotely, via a video call, you would see me represented on your phone's screen as a pixelated image. In it, I'd look like the compound result of tiny rectangular blocks put together. But that doesn't mean that I, Bernardo Kastrup, am made of tiny rectangular blocks. The pixelation is an artifact of my *representation* on a screen, not my inherent structure as that which is represented. This should be obvious enough.

Now, for precisely the same reason, that the structure of a body—a representation of a subject on the screen of perception—is compound, doesn't entail or imply that the subject represented *as the body* is itself compound. Particles are the pixels of the screen of perception—the pixels of the 'physical' world—not necessarily the building blocks of subjects. The structure of the representations isn't necessarily the structure of the represented, and so we cannot conclude that subjects are made of particles; only bodies are.

You may think that these are very abstract and remote considerations, but they constitute the very metaphysical ground where Panpsychism germinated and derives its relevance from. Thus, bear with me.

Unlike particles, cells are living entities just as we are. For this reason, the cellular structure of our brain may seem, intuitively, to be a more compelling sign that our consciousness must itself be compound. After all, there is a certain equivalence between individual cells and our organism as a whole: both are alive and metabolize. As such, if I am conscious, so must the cells that constitute my brain be, and Panpsychism is true—or is it?

Compelling as it may sound, this line of reasoning also relies on an unexamined assumption mistaken for fact. Specifically, when we think of the body as a compound entity just because it is made of many cells, we are conflating *growth* with *assemblage*, and thereby mistakenly taking for granted that our cells are proper parts of us. Allow me to unpack this.

An entity is *assembled* when its structure is defined from the *outside in*, as determined by how its constituent parts are brought together. A car is assembled because its structure is defined by engineers and realized by bonding its parts together in an assembly line. A living organism, on the other hand, isn't assembled; instead, it *grows*. In growth, the structure of the entity is defined from the *inside out*: raw materials still flow in, but their place and role in the organism are defined from within. An organism isn't assembled; its structure is defined by its own inner being.

Only assembled entities can be confidently said to be compound, and thus to have proper parts. Growth, on the other hand, can be coherently interpreted as a process of *inner structuring, differentiation,* or *complexification,* in which the only part is the whole thing. A human being begins life as a zygote—a fertilized egg, a single cell—which complexifies, differentiates, or structures itself internally, in a self-similar or fractal manner, through what we call mitosis (i.e., cell division). *A fully grown person can be coherently regarded as still being the original, non-compound, unitary zygote; just one that has complexified, differentiated, or structured itself internally to a large degree.*

There is, thus, an important sense in which a person—along with her brain—is not 'made of' cells; instead, the cells are simply what the inner complexification, differentiation, or structuring of the person, through growth, *looks like*. Regarding cells as proper parts is, at best, merely nominal.

As such, the cell structure of our body isn't a reason for us to think of ourselves as compound entities, or of our consciousness as made of proper parts. It means only that the *unitary* entity we have always been, since the moment of fecundation, has developed complex and differentiated inner structure over time, through growth. And since that original unitary entity—the zygote—only knew how to be a cell, it shouldn't come as a surprise that it creates inner structure by repeating that original template self-similarly; i.e., in the form of many internal 'cells.' The zygote complexifies itself internally by recursively applying to itself the only template it knows how to be.

To disagree with this is to fail to recognize the reason why we never say that a car grows, or that a person is assembled. The panpsychist intuition would be more apt if a human—or any multicellular organism, for that matter—came into being only when a bazillion individual cells crawled towards each other, and then piled up on top of one another, to form it. But as things stand, it's an intuition based on a failure to understand the difference between assemblage and growth.

Indeed, everything about how our body works tells us that, unlike a car, we *aren't* truly compound entities: the cells in our body 'know' exactly how they need to shape themselves, and what they need to do, depending on where in the body they are located. Their form, activity, and very existence are coordinated by a global, unitary pattern. Moreover, our cells share identical instances of this still-mysterious thing we call DNA, which is the 'physical' clue that they aren't proper parts of a compound entity, but merely the fractal inner structure—complexified through growth—of one irreducible whole. When cells actually behave as parts, we say that they've become cancerous.

In conclusion, not only does Panpsychism have questionable value as a philosophical hypothesis; not only is it flat-out refuted by empirical science; but even the very intuitions that motivate the panpsychist turn out to be based on unexamined assumptions mistaken for facts. Panpsychism isn't the future of metaphysics; it's just the stillborn baby of an attempt to put continuity—that is, the safeguarding of at least *some* aspects of Physicalism—above reason and evidence; the lingering expression of a stubborn *bad habit*. As an alleged remedy, Panpsychism is perhaps worse than the disease—Physicalism—it is meant to cure.

Chapter 6

Analytic Idealism

How peculiar it may seem to you that, well over halfway into a book about Analytic Idealism, I haven't yet explicitly tackled, well, Analytic Idealism! Instead, I've spent most of the foregoing pages discussing the history and flaws of Western thought, Physicalism, Panpsychism, confirmation bias in science, quantum entanglement, the neuroscience of consciousness, our culture's sociopsychology, and a number of other more or less related topics. Does this mean that the core of this book is yet to come? That everything thus far has been merely a kind of context-setting, and that the brunt of the required intellectual effort hasn't even started yet?

No. As a matter of fact, you're almost done already; all the difficult points are already behind us and it's all downhill from here. The reason for it is twofold: first, distributed throughout the previous chapters, often in the form of digressive commentary, are most of the key ideas that underpin Analytic Idealism. By using other metaphysical hypotheses and assumptions as a background to contrast Analytic Idealism with, I've already introduced you to almost all the key tenets of the latter (with one or two exceptions that we will soon tackle). Without even knowing it, if you've followed me this far, you're already almost a graduate in Analytic Idealism.

The second reason why most of the effort is already behind us is even more significant: like an old house that needs to be completely refurbished, most of the required energy has already gone into dismantling the old thought structures and preparing the conceptual ground for the new setup. With that done, bringing the new ideas together, in the form of a cohesive system, is relatively quick and easy.

More explicitly, the main difficulty of gaining a new metaphysical perspective lies in seeing through old, unexamined, unjustified assumptions and (bad) habits of thought. And that we've already accomplished. In and of itself — as I intend to prove to you in this chapter — Analytic Idealism is surprisingly unpretentious and intuitive; almost self-evident. Whatever difficulty may be associated with it is entirely that of disentangling oneself from the tortuous web of abstraction and metaphysical muddle that plagues our culture. When physicalists opine that Analytic Idealism is complex or far-out, they are simply expressing the extent of their own conceptual confusion, and the degree to which they have abstracted themselves away from the concrete reality always present in front of their eyes.

So here is Analytic Idealism for you, in only a few paragraphs.

When we stand on a hill and look out to the horizon, whatever exists beyond the horizon line isn't accessible to us — i.e., it isn't visible. Nonetheless, since we can see the earth up until the horizon, it's natural to think that, beyond the horizon, there is just *more earth*; as opposed to something distinct from, and incommensurable with, the earth. Indeed, if someone were to claim that the earth exists only until the horizon, and that beyond the horizon there is something totally distinct from the earth, you would probably question the sanity of the person.

In just the same way, the analytic idealist inspects everything they have direct access to and finds only *experiential* — i.e., *mental* — states. After all, we are always cooped up in mind. Anything allegedly non-mental, insofar as it can be accessed by us, is *perforce* a mental abstraction of our own mind. Knowledge itself is a set of experiential states. To put this in the language of our analogy, all we can see until the horizon of personal experience is, well, *experience*.

Yet, it's obvious to the analytic idealist—as it is to the physicalist—that there is a world out there, beyond their individual mind; a world *beyond the horizon*, so to speak. However, since all they can see up until the horizon is experiential states, the analytic idealist finds it eminently reasonable to infer that, beyond the horizon, there are just *more experiential states*; i.e., *trans*personal experiential states beyond our personal mentation. And that's why, to the analytic idealist, there really is a world out there, independent of our own minds, yet it is still a *mental*—or *experiential*, terms I use interchangeably—world.

That states beyond our own mind can also be mental is trivial: *my* thoughts are mental and yet external to *your* mind; you cannot access them directly; my thoughts would still exist even if you were not there reading this book; and my thoughts won't change merely because you wish for, or fantasize about, their being different. In exactly the same way, to the analytic idealist nature is constituted of experiential states external to their own mind, which cannot be accessed by the analytic idealist from a first-person perspective, won't cease to exist when the analytic idealist is not observing them, and won't change merely because of the analytic idealist's wishes or fantasies.

In contrast, when the physicalist speculates about what is to be found beyond the horizon of personal, direct experience, they believe it will be something incommensurable with the experiential states they can 'see' up until the horizon; in other words, it's all earth until the horizon, but something totally unlike the earth beyond the horizon. The physicalist mistakes the rather obvious existence of a world beyond *personal* minds for the existence of a *non-mental* world, a complete *non sequitur* (pompous technical jargon for nonsense, baloney).

Under Analytic Idealism, the external experiential states that constitute the outside world *present* themselves to our observation in the form of what we colloquially call the 'physical' world, or 'matter'—i.e., the contents of perception, the stuff we see, hear, smell, taste, and touch. The 'physical' world is merely a dashboard representation of the states of the *real* external world out there, which arise upon our measuring—i.e., observing—the latter's mental states with our sensors: our eyes, ears, tongue, nose, and skin. Just like an airplane represents the output of its sensors in the form of indications on dashboard dials, we represent the output of our sensors in the form of the 'physical' world displayed on the screen of perception. What we colloquially call 'matter' is thus merely an internal cognitive representation of the results of measurements performed on mental states. In other words, *'matter' is what mental states look like when observed from an outside perspective.* And that is all there is to it, in *all* cases, without exception.

Now, because our inner perceptual states—again, the colors we see, the melodies we hear, the flavors we taste, etc.—are also experiential or mental, the whole of nature is mental: perceptual states are *mental* representations of equally *mental* states, and there are no other types of state; all there is is mentation, or experience. Inner perceptual states provide us with information about the external states of the world because they are *modulated by* those external states. And that some mental states can causally modulate other, qualitatively distinct mental states is trivial and known to all of us: our thoughts, despite being qualitatively discernible from our emotions, regularly modulate our emotions and vice versa. Just think of the last time negative emotions led you to pessimistic thoughts, or pessimistic thoughts led you to fearful emotions. So this is how perception works: it's a *mental* process that allows external *mental* states to modulate internal *mental* states. There is only mentation in the whole of nature, without exception. (At this point, all kinds of questions and presumed exceptions must be popping in your mind; bear with me a little longer, for after many years talking to people about these things, chances are I have anticipated your doubts and will clarify them either later in this chapter or in the next one.)

But what, then, is the 'physical' body that seems to enable perception? What are the 'physical' eyes, ears, tongue, nose, and skin that seem to allow us to perceive the world? And what is the brain, that 'physical' organ so closely associated with our ability to think and feel?

Under Analytic Idealism, 'matter'—all 'matter,' without exception—is what mental states *look like* when observed from an outside perspective. In other words, 'matter' is a mental *representation* of other mental states. This applies to the entirety of our 'material' body too: the body is what our inner mentation—including the myriad subtle mental processes below the threshold of metacognitive introspection—*looks like* when represented in the form of perceptual states, through the modulation mechanism discussed above. The body, too, is a mental representation of (our own) mental states. In more traditional language, the body is the *image* of the 'soul' (psyche, mind); it is what the 'soul' looks like when measured from the outside and represented on a dashboard.

Now, more specifically, our 'material' sense organs—eyes, ears, nose, tongue, skin—are what the subset of our mental processes *dedicated to collecting information about the world* looks like, when represented on a dashboard. As Arthur Schopenhauer so evocatively put it over two centuries ago, our eyes are the representation—the image, the appearance—of our *will to see*; our ears, the representation of our *will to hear*; and so on. More generally, our sense organs are what our *will to perceive* looks like, when this will is itself observed and then represented on our internal dashboard of dials. (What is meant by 'will' here is, of course, not our everyday deliberate volition, but something much deeper, instinctive, unpondered, spontaneous to the point of being automatic and inevitable.)

Indeed, if you think of it without metaphysical prejudice, it seems entirely self-evident that our 'material' body is what *we*—i.e., our inner experiences, the only thing we identify with pre-theoretically—look like when observed from the outside. Let me repeat this so its simplicity sinks in: *the living body is what our mental inner life looks like when observed from an outside perspective;* which is to say, our body is what *we* look like from the outside. Put in these words, this is so self-evident it is embarrassing to have to argue for it.

Moreover, since our body is made of the same sorts of fundamental particles or—more accurately—fields as the rest of the universe, it stands to reason that the rest of the universe, too, is what inner experiential states look like when observed from the outside. Matter is the extrinsic appearance of inner mentation when it comes to *both* the matter forming a body *and* the matter forming the inanimate universe as a whole. Could things be any simpler or more consistent?

But the discussion above presupposes some kind of boundary that separates the inside (our personal mentation) from the outside (the world at large). It's because of this boundary that we can speak of an *external* world in relation to our *inner* conscious life. Without it, the inner and the outer would be one and the same cognitive space. Even the notion of dashboard representations presupposes a boundary: sensors make measurements of *external* states, and then display the results on the *internal* dashboard. An airplane is distinguished from the sky by the boundary of its aluminum skin. Therefore, if everything in nature is mental—experiential, qualitative—there must then be *cognitive boundaries* in the mind-space of nature that circumscribe each of our mental inner lives; that define *us* as individual mental agents distinct from the world that surrounds us.

That such boundaries exist is directly implied by the fact that our personal mental lives are *private*: I can't read your thoughts, and presumably you can't read mine either; there is at least one cognitive boundary separating your thoughts from mine. I also don't know what is going on in China right now, let alone in the galaxy of Andromeda; there is at least one cognitive boundary separating me from China and the galaxy of Andromeda. More generally, my cognition is clearly separate from nature-at-large and can only acquire knowledge about the latter insofar as such knowledge is conveyed and mediated by perception. My *individual* mind is not coextensive with the mind of nature-at-large, but constitutes merely a subset of it defined by a cognitive boundary of some sort. It is this boundary that creates the evolutionary impetus—in the Darwinian sense—for sense organs and the internal dashboard of dials that we call our 'screen of perception.' Without the boundary, we wouldn't *perceive* the world, but instead *be* the world and experience it directly, from a first-person perspective.

At first sight, however, mind-space seems devoid of boundaries: I can seamlessly access my memories, thoughts, emotions, fantasies, ideas, etc., as if all these mental contents were part of a unitary database. I don't need to open cognitive doors or pass through cognitive walls to access them. Yet, appearances are deceiving.

Imagine that you are having relationship problems at home. These problems weigh heavily on you, compromising your ability to function normally. You are worried, sad, distressed, afraid, disappointed, angry, etc., all of which disturb the smooth operation of your mind. So when you go to work in the morning, you 'park' your problems—you set them aside in your mind, put them in a 'drawer'—so that you can function. This 'parking' of a part of your mental inner life *is* the creation of a cognitive wall inside your mind. Psychiatrists call it 'dissociation.'

You experience dissociation not only when you park your problems, but also when you forget something that you later recall, and when you experience cognitive dissonance (i.e., holding two mutually contradictory but sincere beliefs at the same time). Dissociation is very common and most of the times not pathological; often it is even a useful psychological defense mechanism. Those of us who suffered childhood traumas, for instance, spontaneously use dissociative mechanisms to protect ourselves against traumatic memories. We 'compartmentalize' our conscious inner lives.

Now the point: every instance of dissociation entails a *cognitive boundary* in mind-space, which is exactly what we were looking for above. Mutually contradictory beliefs can be concurrently held in the same mind because, within that mind, the beliefs are separated by a dissociative boundary. That thing you were trying hard to remember the other day but failed to, was hidden behind a cognitive boundary inside your mind. Sometimes this boundary is highly porous, permeable, and largely under the control of our deliberate volition, such as when we succeed in remembering something, or in parking our relationship problems. But other times the boundary is so opaque and autonomous that our very sense of individual agency—along with certain character traits, specific memories, mannerisms, etc.—splits into multiple, disjoint centers of awareness. When this happens, we speak of an extreme form of dissociation called Dissociative Identity Disorder, or 'DID.'

People with DID present with multiple so-called 'alters,' or 'alternate personalities.' Each alter tends to think of itself as a distinct conscious agent, with its own volition and inner life, separate from the other alters and the host personality. Alters can alternate in taking executive control of the body. Often, they go by unique names and can also claim to have an age and gender different from those of the host personality. They sometimes are aware of the existence of other alters, but sometimes also not. There is significant clinical evidence that different alters can be *concurrently* conscious: they can play tricks on, and undermine, one another, partake of the same dreams concurrently (more on this shortly), and sometimes claim to be aware of the thoughts, plans, or behaviors of other alters as those thoughts, plans, or behaviors unfold (see, e.g., *First Person Plural: Multiple Personality and the Philosophy of Mind*, by Stephen Braude, Routledge, 1995).

Technically, each alter is a segment of the host personality's mind so severely dissociated from the rest that it acquires its own unique vantage point and sense of identity. As such, each alter becomes a distinct center of awareness within what is, otherwise, one single mind-space. The dissociative boundaries that separate alters from one another are thus cognitive boundaries, not physical or extended ones; alters are defined by cognitive isolation, not physical separation in spacetime.

Now, whether we understand the mechanisms of dissociation or not, it is an established empirical fact that dissociation happens in nature and creates boundaries within mind-space. Analytic Idealism leverages this empirical reality by inferring that a *dissociative process* creates the boundary separating our conscious inner life from the rest of nature. In other words, according to Analytic Idealism, *we are all dissociative processes—'alters'—in the one mind of nature.*

The idea is that nature-at-large—which is a mind—undergoes a process analogous to human DID. Just as DID patients present with seemingly disjoint centers of awareness called 'alters,' nature-at-large also presents with multiple disjoint centers of awareness that we call *us*. Analogically speaking, we—and all other living beings—are alters of nature-at-large.

Take a few moments to ponder about this before you read on, so to give yourself time to acclimatize to this fundamental idea underpinning Analytic Idealism.

Now I am going to reframe my argument in field terms, which dovetails nicely with the notion of dissociation discussed above. First, in the next three paragraphs, I will do so briefly and rather intensely, so to give you a sense of overview and make clear how it all comes together. Soon thereafter, however, I will repeat the whole point a lot more slowly, so we can explore its details and underlying intuitions more thoroughly, as well as clarify whatever may have remained unclear the first time round.

Modern physics hopes to, one day, be able to model and predict the behavior of nature with a *single* quantum field, as opposed to the several fields currently postulated in QFT. This effort goes by the name of 'grand unification theories' and there is good reason to believe that it will eventually succeed. When it does, there will be *one* quantum field. Under Analytic Idealism, this one resulting field is, ultimately, a *field of subjectivity, whose particular patterns of excitation —* 'ripples' — *are particular experiential states.* (Strictly speaking, the unified field of physics, insofar as it is extended, will be merely a dashboard representation of the one subject that nature is. But because our entire mode of thinking relies on the paradigm of the dashboard, as opposed to reality, it is easier to pretend that the subject is extended, and therefore can be equated with the unified field of physics.) The one field of grand unification theories is the way we will eventually model the mind of nature. Experiences will be modelled as the field's excitations. As such, we will be able to maintain that there is a sense in which the field *is* the mind of nature.

Now, within this one field, dissociation happens naturally and spontaneously, just as it happens naturally and spontaneously in the mind of a human being. The result of this is, well, *us*; and all other living beings as well. Under Analytic Idealism, *life is what dissociation looks like in the one field of subjectivity that nature is,* when said dissociation is observed from an outside perspective and represented on the dials of a cognitive dashboard.

Granted that what psychiatry calls DID happens only within the minds of *humans*; Analytic Idealism *extrapolates* dissociation 'upwards,' to the next level in the hierarchy of being, and infers that not only does it happen in human minds, but also in the broader, field-like mind-space of nature-at-large. Under Analytic Idealism, human DID is but a self-similar, hierarchically nested micro-reflection, in a human mind, of a broader macro-process that unfolds spontaneously in the one field of subjectivity that nature *is*.

Alright, this was a lot to take in. So let us go over it again, more slowly this time.

Under Analytic Idealism, nature *is* one spatially unbound field of subjectivity. That one field—which corresponds to the unified field of physics—is all there is, there has ever been, and there will ever be. Particular experiential states are merely particular *patterns of excitation*, or 'ripples,' of this one natural subject. Each different pattern of excitation is thus a discernible, countable experiential state. That's how the unfathomable variety and complexity of natural states is accounted for in terms of the simplicity of *one* field of subjectivity: just as a single guitar string can oscillate differently so to produce many different notes, the one subject of nature can 'oscillate' differently so to produce what is represented on our cognitive dashboard as blackholes, quasars, pulsars, nebulae, galaxies, stars, planets, moons, mountains, volcanoes, oceans, etc.

Notice that the inferential move I'm using here is entirely analogous to the one physicists use to account for complexity in terms of simple fields. There is nothing new or peculiar about the line of thinking I am developing; on the contrary: it is a move that has proven extraordinarily productive in science since Michael Faraday proposed electromagnetic fields in the 19th century.

Now, in addition to 'rippling,' the one field of subjectivity that nature is can also form 'whirlpools': it can configure itself, spontaneously, so to form highly localized patterns of excitation that turn in upon themselves and become seemingly distinct from their surroundings. This is how we can visualize dissociation over a field: a particular alter in the mind of nature is akin to a 'whirlpool' in the 'river' of natural mentation.

When we find a whirlpool in a river, we can point at it and say, "There's a whirlpool!" We can precisely determine its location and delineate its boundaries. Yet, there is absolutely nothing to a whirlpool but the river, in movement. The whirlpool is not a thing, but a doing of the river, this being the reason why we can't lift the whirlpool out of the river and take it home with us. In precisely the same way, there is nothing to an alter of the mind of nature but, well, the mind of nature. The alter is not a thing, but a *doing* of nature; and this particular kind of doing is what we call a *dissociative process*. Yet, we can point at an alter of nature and say, "There's a person!" We can precisely determine its location and delineate its boundaries, just like a whirlpool in a river.

As such, the *only* irreducible natural entity under Analytic Idealism is the one field of subjectivity that nature is. Everything else is the configurations and patterns of excitation of this one field, which—for being mere *behaviors* of the field—are all reducible to the field. There is nothing but the field, just as there is nothing to ripples and whirlpools but the water that ripples and turns. Analytic Idealism is not only the most parsimonious metaphysics there is, but also the most parsimonious *there can be*; for any coherent theory of reality has to propose at least *one* irreducible entity. And when that one entity is the one empirical *given* of nature—i.e., subjectivity itself, which is where all theorizing unfolds—the theory is then the most parsimonious possible; it postulates nothing beyond the *one given fact* of reality: the existence of subjectivity.

When the analytic idealist speaks of a field of subjectivity 'underlying matter,' this shouldn't be taken to mean that two irreducible entities are being proposed (namely, the field of subjectivity *and* matter). Doing so is an uncharitable and rather thick misunderstanding that, unfortunately, even professional philosophers and university professors are known to have been guilty of. One must remember that the analytic idealist isn't communicating to the culture at large with the same constipated conceptual rigor with which academic papers are written. By 'underlying matter,' the analytic idealist means that what we call 'matter' is but a discernible *appearance* or *representation*, on our inner cognitive dashboard, of patterns of excitation of the field of subjectivity. *Only the field of subjectivity exists;* even the inner representations are themselves patterns of excitation of that one field, since an alter is still a segment of the field just as a whirlpool is still a segment of the river.

Another terminology trap to avoid is the following: when the analytic idealist says that experience is fundamental, their point is that the *one field of subjectivity* is fundamental; for experiences are but excitations *of* the field, there being nothing to experience but the field (just as there is nothing to a ripple but the river). It is in this sense that the analytic idealist may use the words 'experience,' 'mentation,' 'mind,' 'subjectivity,' and 'subject' interchangeably: in all cases, there is nothing but the subject (or the mind, the field, whatever one wants to call it). The analytic idealist does not differentiate experience from the experien*cer*; there is *only* the universal experien*cer*, all experiences being but patterns of excitation *of* the universal experiencer.

Let us now look more carefully into how perception works under the model I've just outlined. Each alter in the mind of nature—i.e., each living organism—is defined by its respective dissociative boundary. Experiential states *within* the dissociative boundary constitute the alter's private conscious inner life. Experiential states *outside* the dissociative boundary comprise the transpersonal experiential states that constitute the external world of the alter. But these external states can still *impinge* on the dissociative boundary from the outside-in; they can 'bump against' the boundary, so to speak, thereby influencing the dynamics of the alter's inner experiential states.

To see how this works, let us return to the analogy of 'parking' your relationship problems when you go to work in the morning. Notice that, while you are at work, although the disturbing emotional states that you've parked remain unexperienced by your ego, they still impinge on your ego's functioning: they may color your thinking negatively, leading you to more pessimistic assessments of situations; they may trigger your temper, leading you to react in an unfriendly manner to social circumstances; they may bias your cognitive associations and memory access; etc. What this shows is that, even when certain experiences are dissociated from the executive ego—in the sense that they aren't experienced by the ego—they may still influence, 'from the outside,' what the ego *does* experience. I shall call this causal influence, extending across a dissociative boundary, an 'impingement.'

Under Analytic Idealism, as we've discussed above, what we call life—biology, metabolism—is nothing but the extrinsic appearance of a dissociative process in the field of subjectivity that nature is; a dashboard representation of the dissociation. Science tells us that life arose in a simple form, probably akin to a procaryote or an archeon (bacteria-like single-celled organisms), about four billion years ago. The defining characteristic of single-celled organisms is the cell membrane that separates them from their environment. Under Analytic Idealism, that membrane is what the organism's dissociative boundary looks like when observed and represented on a dashboard. Equipped only with this membrane as a means to interact with their environment, the first living organisms could collect information about their surroundings only in a very vague and rudimentary way, as they had no specialized sense organs. All they could sense from across their dissociative boundary were stimuli that contacted their membrane, in a manner somewhat akin to touch or taste, but much less clear and detailed. It was only over time that sense organs evolved, which could tune into, pick out, and amplify these stimuli, so to convey a more detailed picture of the surrounding environment.

Let me now tell this same story again, but this time *not* in the language of dashboard representations—i.e., the language of the 'physical' world of cells and membranes—but in the language of *reality*, of the thing in itself—i.e., alters and dissociative boundaries. The first alters had inner experiential states very poorly modulated by external states. Think of it as the impingement of your parked emotions while you are at work: while real and noticeable, it is also very limited and unclear; it provides just a hint of what you are repressing, not a clear picture of it. This is comparable to how photons or air pressure waves bumping against the skin of your hand, despite imparting measurable energy on it, are hardly registered in your conscious inner life.

But the wonderful thing about evolution is that, given enough time, it finds ways to zero-in, isolate, and amplify whatever environmental information inflow is conducive to survival, even the faintest. And so it was that life started forming tissues and systems dedicated to tuning into, and picking out, those extremely subtle impingement stimuli, since they convey useful information. When photons now bump against your *retina*—instead of the skin of your hand, or the membrane of a single-celled organism—the same energy is picked out, amplified, and thereby experienced in a much richer and detailed manner: vision. Similarly, when air pressure waves bump against your *eardrums*, that same energy is picked out, amplified, and thereby experienced in a much richer and detailed manner: hearing.

Your parked emotions don't convey much discernible information when they impinge against your ego because you didn't evolve to pick them out. There is no survival advantage to collecting detailed information about mental processes that are dissociated from your ego precisely so you can be *more functional*. But when those mental processes constitute the world around you, then it is indeed extraordinarily useful to tune into their impingement. Perception is the result of this evolutionary pressure: our need to pick out otherwise faint and vague impingements on our dissociative boundary.

Perception provides us with an at-a-glance view of the experiential environment around us in an encoded form that not only caps our internal entropy, but also highlights whatever is salient about the world, so we can react more promptly to environmental challenges and opportunities. Indeed, this is the reason why perceptual states are so *qualitatively different* from endogenous experiences such as thoughts and emotions. If we were to collect information about our environment in a manner qualitatively similar to the environment as it is in itself, we would be like skilled telepaths in the middle of an agitated crowd: there would be no sense of overview, just confusion. But when we contemplate that crowd through 'physical' vision and hearing, we have an immediate overview of what's going on, which allows us to react more promptly and effectively to both challenges and opportunities.

This, in a nutshell, is the account of perception under Analytic Idealism. And it is one that avoids the 'hard problem of consciousness' altogether: the qualities of perception are modulated by *other* qualities—namely, the experiential states that constitute the world around us—instead of being somehow generated by non-qualitative, physical stuff. A magical bridge from quantities to qualities is no longer required; only a causally trivial one, from qualities to *other* qualities.

But if you are a critical reader, by now you have a list of questions and points of skepticism about my narrative. So let us try to tackle some of them.

I've been maintaining that living organisms are merely *extrinsic appearances, representations* of dissociative processes in nature-at-large. But since my argument is based on an analogy with DID, this raises a question: do the dissociative processes in the mind of a DID patient have extrinsic appearances too? Do they look like something when the patient's brain is scanned?

Yolanda Schlumpf and her team, in the Netherlands (see: "Dissociative part-dependent resting-state activity in Dissociative Identity Disorder: A controlled fMRI perfusion study," published in *PLoS ONE*, 2014), performed functional brain scans on both DID patients and actors simulating DID. The scans of the actual patients displayed clear differences when compared to those of the actors, showing that dissociation indeed has an identifiable extrinsic appearance. In other words, there is something rather particular that dissociative processes *look like* from the outside. This substantiates the notion that living organisms, such as you and me, are what dissociative processes in nature-at-large look like. Metabolizing bodies are to dissociation in nature-at-large as certain patterns of brain activity are to DID patients. It's just that, in the case of nature-at-large, we don't need brain scanners, since the 'brain' in question—i.e., the dashboard representation of the mind-space in question—is the universe in which we are already immersed.

This immediately raises another question: if the dissociative processes of a DID patient unfold within a brain, why does the 'physical' universe within which life—also a dissociative process—unfolds look nothing like a brain?

Well, does it really *not*?

As it turns out, the network topology of the universe at its largest scales does resemble that of a brain; so much so that astrophysicist Franco Vazza and neuroscientist Alberto Feletti considered the similarity "truly remarkable" and "striking" (see: "The strange similarity of neuron and galaxy networks: Your life's memories could, in principle, be stored in the universe's structure," published in *Nautilus*, on July 20, 2017). They say:

> It is truly a remarkable fact that the cosmic web is more similar to the human brain than it is to the interior of a galaxy; or that the neuronal network is more similar to the cosmic web than it is to the interior of a neuronal body. Despite extraordinary differences in substrate, physical mechanisms, and size, the human neuronal network and the cosmic web of galaxies, when considered with the tools of information theory, are strikingly similar.

Now, of course the cosmic web and an organic brain are not identical; as Vazza and Feletti concede, there are obvious differences in scale, composition, and function. But this shouldn't be surprising either: dissociation at the level of nature-at-large is *not* identical to human DID, for equally obvious reasons. The two processes are posited to be just quite *analogous*. And that strong analogy is consistent with the strong topological similarity between the cosmic web of galaxies and the organic brain. As a matter of fact, this topological similarity is wholly incomprehensible—there is nothing in the known laws of physics that could even remotely account for it—without the explanatory framework of Analytic Idealism.

But truly critical readers will still be far from content. They will reason that the alters in the mind of a human with DID cannot see, touch, or otherwise interact with one another, the way you and I can see each other and shake hands in the broader context of the world we share.

But then again, is this *really* true?

To answer this question, the first thing we must remember is that, unlike the case of a human with DID, there is no external world from the point of view of nature-at-large. After all, the latter is all there is. So to make a fair comparison, we must compare the experiential inner life of nature-at-large to a human's *dream life,* for only then all experiential states in both cases are internally generated, without the influence of an outside world.

What do we know, then, about the dream life of a human DID patient? Can the patient's different alters share a dream, taking different co-conscious points of view within the dream, just like you and I share a world? Can they perceive and interact with one another within their shared dream, just as people can perceive and interact with one another within their shared environment? As it turns out, in a large minority of DID patients this is precisely what is reported, as clinical research done at Harvard University has shown (see: "Dreams in dissociative disorders," by Deirdre Barrett, published in *Dreaming,* 1994). Here is an illustrative segment from a dream report of one of the subjects studied:

The host personality, Sarah, remembered only that her dream from the previous night involved hearing a girl screaming for help. Alter Annie, age four, remembered a nightmare of being tied down naked and unable to cry out as a man began to cut her vagina. Ann, age nine, dreamed of watching this scene and screaming desperately for help (apparently the voice in the host's dream). Teenage Jo dreamed of coming upon this scene and clubbing the little girl's attacker over the head; in her dream he fell to the ground dead and she left. In the dreams of Ann and Annie, the teenager with the club appeared, struck the man to the ground but he arose and renewed his attack again. Four year old Sally dreamed of playing with her dolls happily and nothing else. Both Annie and Ann reported a little girl playing obliviously in the corner of the room in their dreams. Although there was no definite abuser-identified alter manifesting at this time, the presence at times of a hallucinated voice similar to Sarah's uncle suggested there might be yet another alter experiencing the dream from the attacker's vantage. (page 171)

This shows that, while dreaming, a dissociated human mind can manifest multiple, concurrently conscious alters that share the dream and experience each other from second- and third-person perspectives, just as you and I can shake hands with one another in ordinary waking life. The alters' experiences are also mutually consistent, in the sense that all alters seem to perceive the same series of events, each alter from its own individual vantage point. The correspondences with the experiences of people sharing an outside 'physical' world are self-evident: alters in the mind of a human can not only see one another, they can even club each other over the head!

Nonetheless, you may remain skeptical that dissociation can be strong to the point of blinding us to each other's thoughts and feelings, given that we are all ultimately supposed to be—as Analytic Idealism maintains—in the same mind-space. Yes, the alters of a human DID patient can lose access to some of the host's memories and all, but our predicament seems to be of a whole other order of magnitude. We are *very* alienated from each other's inner lives, let alone from the inner lives of other species. Lack of empathy is perhaps humanity's greatest—and most dangerous—shortcoming. We are always busy waging cruel, terribly destructive wars against each other and the planet as a whole. Can dissociation be *that* thorough?

In 2015, doctors in Germany (see "Sight and blindness in the same person: Gating in the visual system," by Hans Strasburger and Bruno Waldvogel, published in *PsyCh Journal*) reported the extraordinary case of a woman who presented with a variety of alters. Peculiarly, some of her alters claimed to be blind. Using an EEG instrument, the doctors were able to ascertain that activity in the visual cortex—the part of the brain associated with sight—indeed wasn't present while a blind alter had executive control, even though the woman's eyes were wide open. Remarkably, when a sighted alter or the host personality assumed control, the usual activity in the visual cortex returned.

This compellingly demonstrates the *literally blinding* power of dissociation. If it can make you literally blind to what is right in front of your open and healthy eyes, what *can't* it make you blind to? In this context, it isn't at all a stretch to imagine that dissociation blinds us to each other's thoughts, and to what is happening in the galaxy of Andromeda, thereby inculcating an illusory and limiting sense of personal identity.

Clearly, thus, our empirical grasp of dissociation shows that a DID-like process at a universal scale is, at least in principle, a viable explanation for how individual minds—represented as living organisms on the dashboard—arise within the one field of subjectivity that nature is. Whether the precise mechanisms of dissociation are also understood or not—something I will speculate about in a later chapter—is but a secondary question: whatever these mechanisms may be, we know empirically that they do exist in nature and produce precisely the effects we would expect them to produce, if biology indeed is but the extrinsic appearance of alters in nature-at-large.

As discussed earlier, under Analytic Idealism all 'matter' is merely a dashboard representation of experiential states. More specifically, the living body, insofar as it is 'made of matter,' is a dashboard representation of the experiential states *within an alter* of nature-at-large. You and I are examples of such alters. Therefore, our bodies are dashboard representations of our dissociated, internal experiential states.

But this seems to pose a problem: what experiential states does, say, your *liver* represent? Or your left big toe? It's easy to accept that active parts of your brain correspond to your personal experiences; indeed, they are called the 'Neural Correlates of Consciousness,' or NCCs. But it's more difficult to find, through introspection, an experience that presents itself as your femur, or your kidneys, or your appendix, isn't it? We're used to thinking of the nervous system as something intimately associated with our conscious inner life in some way, but not the rest of the body. Yet, according to Analytic Idealism, *every* part of our body represents experiential states, for that is what the body *is*: a 'physical,' dashboard representation of the mental contents of an alter.

To understand this, the first thing we have to realize is that there is a large difference between having an experience and being able to *report* that experience, to others or to oneself. The reportability of experience is an extra property that comes *on top of* the experience itself; it isn't entailed or implied by the latter.

For instance: we've been talking a lot about dissociation. We've also discussed the hypothesis of a self-similar hierarchy of dissociation in nature. At a first level in the hierarchy, the field of subjectivity that nature is dissociates into livings beings, such as you and me. But you and I can also further dissociate—a second level of dissociation—into DID alters. And those DID alters can still further dissociate—a third level—when they forget something, 'park' their troubles, or enter states of cognitive dissonance.

The point I am trying to make is that human minds, even when they are not suffering from pathological forms of dissociation such as DID, naturally dissociate into multiple mental complexes, the most prominent of which is the ever-present executive *ego*. And the ego is the complex that *reports* experience, to others and to itself. Yet, it obviously cannot report the experiences of other complexes, for those are, by hypothesis, dissociated from—and therefore not accessible to—the ego. The mental states represented on a cognitive dashboard as your liver, kidneys, appendix, and left big toe, are dissociated from your ego.

Indeed, there is no evolutionary advantage—only disadvantages—to the executive ego having associative access to states corresponding to *autonomous* functions such as heart rhythm, blood filtering, digestion, structural support, etc.; they require no deliberation or decision-making—as they must always run anyway—and are best left to themselves. To see this, just try to take explicit egoic control of every muscle movement you make while riding a bike; see how long it will take you to decide that it's best to leave those muscle movements to themselves. Everything about ourselves that we refer to as an 'automatism'—be it a physiological automatism such as kidney function, or a function that has become automated through training, such a riding a bike—is a mental complex dissociated from the executive ego to some degree.

You—i.e., your ego—cannot access the mental states corresponding to most autonomous functions, for they are naturally dissociated from you by construction. That's why you cannot introspectively find the experiences corresponding to your liver, or your left big toe. This is not a far-fetched hypothesis at all, much to the contrary: any therapist knows that most of our inner life is indeed dissociated from our ego, this being the reason why people often need years of therapy to acquaint themselves with *some* of what they are *really* thinking, feeling, and generally experiencing at all times.

Another reason why experience and reportability don't always come hand-in-hand is that reportability requires a higher-level mental function called *metacognition*. Metacognition is our ability to shift our attention, introspectively, to the contents of our own mind, thereby inspecting, evaluating, or otherwise taking explicit notice of them. When we do this, we *re-represent* our own mental states: we mirror raw experience on higher-level re-representations. These re-representations are themselves also experiences: the experiential *knowing of* another experience. It is the re-representation—the knowing of another experience—that we report to ourselves and others when claiming that we're experiencing this, that, or the other thing. Without the re-representation nothing can be reported, for we won't know *that* we are experiencing.

Jonathan Schooler published a seminal paper in 2002 (see: "Re-representing consciousness: dissociations between experience and meta-consciousness," published in *Trends in Cognitive Science*) showing the distinction between raw experience—or representation—and their metacognitive re-representations. The contents of consciousness that we can report to ourselves correspond to *meta*-consciousness—the metacognitive, re-represented subset of consciousness—not raw consciousness.

To report an experience to ourselves, *in addition to* having the experience we *also* need to re-represent it metacognitively. Only then do we know *that* we are having the experience. Therefore, when an experience goes unreported it either isn't happening or isn't re-represented, because we can't—or won't—access it metacognitively. For instance, most of the time we don't bother to re-represent the experience of breathing: the expansion and contraction of our ribcage, the air flowing in and out of our nostrils, etc. Yet, we are experiencing our breath at all times. Also, some people—particularly men, in my experience—often fail at re-representing mild forms of pain. I am one of them. I can go for days feeling pain, but without reporting it to myself; i.e., without knowing *that* I am feeling the pain. If you were to ask me, "Bernardo, are you in pain?" I could easily—and sincerely—tell you that I am not, just to retrospectively realize, days later, that I have actually been in pain for quite a while. My inability to report my own pain to myself is explained by the fact that I often fail to re-represent my pain.

There are even medical conditions in which the experiences we tend to be most meta-cognizant of, such as vision, can fall outside the field of meta-consciousness. 'Blindsight' is one of these conditions. People with blindsight claim to be unable to see, e.g., the movement of a ball, but react to that movement just as though they *did* see it. These people do experience the seeing of the ball—as their behavior betrays—but cannot report, to others or themselves, *that* they experience it; they have the experience of seeing the ball, but are not aware *that* they see the ball, due to a lack of re-representation (see, e.g., "The source of consciousness," by Ken Paller and Satoru Suzuki, published in *Trends in Cognitive Sciences*, 2014).

Because of internal dissociation and the limited scope of meta-consciousness, the ego cannot report most of our internal mental states. And that is why we cannot explicitly introspect into the states corresponding to our liver or left big toe: they simply fall beyond the reach of egoic meta-consciousness, despite being experiential, mental, qualitative, subjective states. The much-lauded NCCs are merely the neural correlates of egoic meta-consciousness, not phenomenal consciousness in general. All other neuronal processes are still phenomenally conscious— or, more accurately, they *are consciousness*—and so are the non-neuronal processes that constitute the rest of the body.

There is now just one more point we need to cover before wrapping up this brief overview. In the previous chapter, I criticized Panpsychism for simply adding one more fundamental element to the reduction base of Physicalism, instead of explaining anything. It imports all of Physicalism and then proclaims, "Oh, and by the way, experience is just one more irreducible property/essence of matter." This is a subterfuge to avoid the need for an explanation, not an account of anything.

Nonetheless, Analytic Idealism *also* considers experience— i.e., felt qualities, subjectivity, mentation—irreducible. Isn't this move also a subterfuge to avoid the need for an explanation? No. And the reason is that, unlike Panpsychism, Analytic Idealism does *not* import the reduction base—the set of irreducible, fundamental entities—that Physicalism adopts.

Because we can't keep on explaining one thing in terms of another forever, all theories of reality must have at least one fundamental entity in its reduction base. The better theories are those that can explain *everything else* in terms of that one irreducible entity. The *best* theory chooses, as the one irreducible entity, nature's sole pre-theoretical given: subjectivity itself, of which experience is just an excitation. When placing transpersonal subjectivity into the reduction base, Analytic Idealism *removes everything else from it*; it removes fundamental particles and their physical properties from it, for these particles and associated properties can now be explained *in terms of* excitations of transpersonal subjectivity. This isn't just an effective explanatory move, it is also the best possible explanatory move.

Panpsychism, on the other hand, removes precisely *nothing* from the reduction base of Physicalism. Worse yet, it in fact creates a straw man of Physicalism by mistakenly presupposing that bazillions of discrete fundamental particles—little 'marbles'—dispersed across the universe are all irreducible (an adequate understanding of modern Physicalism entails 'only' about 17 different quantum fields in the reduction base). This is why Panpsychism explains *nothing*: it only adds more 'brute facts' to our epistemology, making sense of none. Analytic Idealism, in turn, removes *everything* that Physicalism added to the reduction base. Its explanatory power is *maximum*: in principle, it accounts for *everything* in terms of just *one* entity, that one entity being the only empirical given we have.

This completes our brief but fairly thorough overview of Analytic Idealism. Nonetheless, because the idealist perspective is so different from what you have been told or absorbed by cultural osmosis throughout your life, I must help you to slowly get more acclimatized with it. This is the purpose of the next chapter, which won't introduce any new element to the story I already told you, but will hold your hand as we try, together, to apply this new understanding to interpreting the myriad facets of empirical reality. In the process of doing so, I may— and probably will—also address other questions or points of skepticism that may have arisen in your mind as you made your way through the foregoing.

Chapter 7

Circumambulation

'Circumambulation' is a powerful Jungian concept. To circumambulate something is to walk round about it. In its epistemic sense, to circumambulate means to approach a theory or idea from multiple different perspectives or angles, taking notice of as many of its facets as possible, each time under a different 'light,' so to speak. Unlike the traditional linear approach, in which a topic is systematically covered from beginning to end, without repetition, circumambulation entails multiple returns, a constant revisiting of subtopics already covered, but in a different context, a different order, from a slightly different slant.

Indeed, you may have noticed that, while a linear structure predominates in the foregoing chapters, I already discretely used some circumambulation there too, regularly returning to material already covered, but each time in a different context. I even already hinted at circumambulation itself—without defining it—earlier, trying to plant the idea in your mind so to give you time to wonder about it before confronting it more explicitly. When the theory one is contemplating violates many of our most ingrained—though unexamined—assumptions, only circumambulation can acquaint us with it in a minimally thorough and self-consistent manner. So this is what we are going to do in this chapter, but this time more explicitly and almost without linear structure.

Here we go.

Even if you agree with the ideas presented in the previous chapter, now and then scenarios will pop into your mind that will seem to contradict Analytic Idealism. One scenario in particular is quite common. It goes like this: "Bernardo, I understand that, according to Analytic Idealism, mind is fundamental. But when I drink a glass of wine, something changes in my awareness. And if I prick your arm with a needle, something happens in your awareness. In both cases, a physical cause produces a mental effect, so clearly mind is derivative, caused by physical stuff!"

While obvious, the error here may evade you because of our deeply ingrained habits of thought: the notion that a physical cause produces a mental effect presupposes some form of *Dualism*, the view that there are physical things distinct from mental things. But Analytic Idealism denies Dualism. All 'physical' causes—such as the glass of wine you drink and the needle you plan to stick into my arm—are merely *dashboard representations of mental states*. In this particular case, the mental states in question are *transpersonal* states out there in the world, beyond our dissociative boundary: the wine is what a transpersonal set of mental states looks like when we observe and represent it on our inner dashboard; the needle, too, is merely what a transpersonal set of mental states looks like, when observed from across our dissociative boundary. 'Matter'—*all* 'matter,' without exception—is merely what mental states look like from across a dissociative boundary.

When you drink the wine or pierce your skin with the needle—let's keep *my* skin out of this—you bring the external mental states 'underlying' the wine and the needle *into your alter,* making them cross your dissociative boundary; you 'absorb' them, so to speak. Once inside your alter, they causally react with your own internal mental states, leading to the experiences of pain and drunkenness. Now, that a mental process in your alter can interfere with other mental processes also in your alter is trivial: your thoughts interfere with your emotions, and vice versa, every day, without raising any metaphysical eyebrow. Is it then counterintuitive that, when you bring an external mental process into your alter, you experience internal mental changes? The causality here goes from mental to mental, not from physical to mental, for all seemingly physical things are merely 'physical' things.

We are thoroughly indoctrinated into thinking of everything that appears on the screen of perception as non-mental, physical stuff: drinks, pills, needles, surgical scalpels, electrodes, etc. But this is merely a tenet of Physicalism, not an empirical fact. To properly comprehend Analytic Idealism without begging the question—whether you ultimately agree with Idealism or not—you must part with this bad habit. Admittedly, it takes discipline and time to kick it off, but it must be done; for under Analytic Idealism, all 'physical' things are just appearances, representations of mental states. That 'physical' things have mental effects only means that the 'underlying' mental processes can causally interfere with other mental processes, which is trivial. All causality is mental-to-mental, even when the mental states involved are 'physically' represented.

"If the world out there is mental, made of qualities, does it mean that the colors I see, the flavors I taste, the smells I feel, etc., are really out there in the world?"

No, it doesn't mean that at all. The qualities of your perception exist only within the dissociative boundary of your alter. In other words, they are inside your personal mind—though not in your head, as we discussed earlier—for they are representations displayed on your inner dashboard. But what they represent is also mental, even though not consisting of the qualities of your perception. For example: the perceived redness of an apple exists only in the mind of an observing alter; it is not external. But the redness perceived represents *another quality*—which isn't itself redness—out there in the world. For as long as we are alive, we cannot access that external quality directly, from a first-person point of view, because we are dissociated from it; life *is* the dissociation that separates us from the qualities of the external world as they are in themselves. All we have to collect information about our surroundings are the qualities of perception, which are internal to us but represent external qualities, just as a dashboard represents external states without being itself external.

"Under this view, what are the elementary subatomic particles we detect in the external world?"

Elementary particles are akin to the 'pixels' *of the screen of perception*, not the fundamental building blocks of the real external world. The particles—i.e., the 'pixels'—correspond to the smallest excitations of the field of subjectivity surrounding us that we can still discern through perception, potentially with the use of instrumentation. The latter—such as microscopes and particle accelerators—amplify the mental excitations constituting the world around us before presenting them to us *still through perception* (after all, we must *perceive* the outputs of instruments for them to be of any use to us). Sometimes we don't even need instrumentation to pick out the lowest-energy discernible excitations that impinge on our dissociative boundary: under the right conditions, we can, for instance, see individual photons as they impinge on our retina. Photons are thus the pixels of visual experience. They and other elementary particles define the structure of our internal dashboard—i.e., the 'physical' world—not that of the real external world. This is analogous to how the pixels constituting my image on your phone's screen define the structure *of the screen*, not that of Bernardo Kastrup.

"But when you talk of photons impinging on our retina, you are implicitly recognizing that there are physical photons out there, as well as physical retinas!"

No, I am not. There are always two ways to speak of anything in nature, two different 'languages': one is the language of the *representations* (i.e., of things as they appear on the screen of perception) and the other is the language of the *things in themselves* (i.e., the *real* world, things as they are 'before' they are represented, or 'behind' the representations). When I speak of "photons as they impinge on our retina" I am using the language of representations. In the language of the things in themselves, I could have said: "external mental states as they impinge on a segment of a dissociative boundary related to visual experience." These are two alternative ways to say the same thing; one implies the other.

But why did I choose the language of representations not only above, but also in several other passages throughout this book? The reason is that our culture is indoctrinated into thinking that the dashboard *is* the world; that the representations are the thing represented. Therefore, we only use the language of representations, and all our semantic references are grounded in that language. Insofar as I want to be understood, I must thus appeal to our shared semantic references, which often forces me into using the language of representations. My doing so, however, should not be construed as an acknowledgment of physicalist premises.

"I don't know, Bernardo. The physicalist account of perception adds up so perfectly. Photons hit our retina and we see them, but they can also hit a photodiode and be detected, even counted, by the diode. Measurable air pressure oscillations can be measured by instruments designed to detect them—such as microphones—but they can also be heard by us, in a way that is totally consistent with the microphones' measurements. The physicalist account of perception just adds up!"

The above has absolutely *nothing* to do with Physicalism as a metaphysics. What the intuition above means is merely that *perception is an internally consistent process*: the direct *perception* of photons is consistent with the *perception* of the behavior of electronic photodetectors; the *perception* of the outputs of instruments measuring air pressure waves is consistent with the direct *perception* of sound; and so forth. And of course perception is self-consistent: the dashboard evolved to be so, otherwise it would just confuse us.

What we must realize is that photons and air pressure waves impacting on retinas and eardrums are *perceptual representations of the process of perception*, not the process of perception as it is in itself; for the only objective access we *can* have to the process of perception is by, well, *perceiving* it. We *perceive perception*, and the result of that is photons and air pressure waves impacting on retinas and eardrums. Photons, air pressure waves, retinas, and eardrums are themselves perceptual representations; what else?

Dashboards can represent the very process of dashboards representing stuff. And when they do so, the results will *perforce* be self-consistent, because we are dealing with an obvious recursion here: the dashboard can do nothing other than to impose its own representational paradigm on its representations of itself, so it all adds up beautifully. This has nothing to do with Physicalism. The self-consistency of perception perceiving perception is simply an unavoidable feature of the recursion entailed; you get it *by construction*.

Let me try to put it as simply as I can: *perception will always represent perception in a manner that is consistent with, well, perception.* Is this surprising? Is it logically sound to construe this tautology as evidence for Physicalism?

"Okay, but how do I fit physics into all this? I mean, the equations and theoretical entities with which we model nature? They don't seem to model mental states, and yet they work!"

Nature, as it is in itself, is constituted by mental states not amenable to description through physical quantities. What is the length of a thought in inches? The weight of an emotion in pounds? Now, these *real* mental states can be represented on a dashboard as 'physical' states: colors, sounds, smells, flavors, and textures. These representational 'physical' states, in turn, *can* be described through physical quantities, for physical quantities were invented precisely to describe them.

Physics models nature through physical quantities. Therefore, it describes and predicts the behavior *of the dials on the dashboard*. But since the dials are constructed so to convey salient information about the world outside, *indirectly* physics models the world: it describes and predicts the states of nature *through the intermediation of their representations on a dashboard.*

So physics does work; it does what it thinks it does, just not in the *direct* way it thinks it does it. Nonetheless, as far as practical applications are concerned, this difference is of little consequence, as we are anyway always limited to the dashboard in our interactions with nature.

In summary: the equations of physics describe the dynamics of the states of the dashboard, which in turn represent the real states of the world.

"If, according to Analytic Idealism, the body is what our inner conscious states look like from the outside, why doesn't the body disappear when we pass out and become unconscious? Or during dreamless sleep and general anesthesia?"

Even when the reporting ego goes offline, the myriad other mental states that aren't reportable by the ego—because they are dissociated from it or lack metacognition—remain unaffected. And those correspond to the body, while the ego corresponds only to particular patterns of brain activity—the 'NCCs,' or Neural Correlates of Consciousness—which can temporarily disappear.

Research has shown that, even when we are in dreamless sleep, we still have experiences that fall into three broad categories: sleep-perception, states of no-self, and sleep-thinking (see: "Does consciousness disappear in dreamless sleep?" by Jennifer Windt *et al.*, published in *Trends in Cognitive Sciences*, 2016). We just don't recall these experiences when we wake up, just as we often don't recall our—very experiential—dreams. There is also overwhelming evidence that when we pass out—which is technically called 'syncope'—we have surprisingly rich experiences. Teenagers worldwide have figured this out, and so engage in a dangerous activity called 'the choking game': they subject themselves to partial strangulation to pass out and essentially have a psychedelic trip without drugs. Pilots that pass out during G-force training also report "memorable dreams" (see: "Acceleration-induced loss of consciousness: A review of 500 episodes," by J.E. Whinnery and A.M. Whinnery, published in *Archives of Neurology*, 1990). There is also significant evidence for forms of experience during general anesthesia (see, for instance: "Anesthesia and Consciousness," by John Kihlstrom and Randall Cork, published in *The Blackwell Companion to Consciousness*, 2007). Indeed, one of the drugs in the general anesthesia cocktail is meant precisely to prevent the formation of memory pathways, so patients cannot later remember their experiences during the period of unresponsiveness; when they wake up, they think they experienced nothing. This doesn't mean that patients experience pain during surgery—the latter would correlate with elevated heart rate, blood pressure, and other indicators that anesthesiologists constantly monitor—but that they have some form of experience they can't later recall.

The evidence thus indicates that what we take to be states of unconsciousness are, in fact, merely states of *unresponsiveness*. We are never unconscious; we simply cannot report—not even to ourselves—many of our experiences, because of a lack of metacognition, memory pathway formation, or associative links to the reporting ego.

"Bernardo, if you think everything is consciousness and there are no physical things, try to step in front of a very physical speeding truck!"

Touché! You got me! No, not really, you didn't. To say that everything is mental is *not* to withdraw the causal powers of things. It is only under the premises of Physicalism that mental stuff—for being epiphenomenal—is kind of meek, causally neutered, or even nonexistent (cf. the insanity of Eliminativism and Illusionism, which I will not waste your time talking about here). Under Analytic Idealism, in turn, *mental states are the carriers of all causal powers;* even those mental states that look like a truck coming your way.

There are transpersonal mental processes out there that can impinge on your dissociative boundary so strongly, they put an end to the dissociation that goes by your name—i.e., they kill you. Think of it as a wave coming up the river and then hitting a whirlpool upstream so hard that it disrupts and dissolves the whirlpool. When you observe that incoming wave from across your dissociative boundary—i.e., from inside the whirlpool—and represent it on your dashboard of dials, it may very well look like a truck coming your way. If you want the alter that you are to keep on going—i.e., to remain dissociated from the rest of nature, to stay alive—take the truck seriously, just not literally.

Any pilot flying by instruments alone knows better than to take the dial indications on their dashboard lightly. Yes, the dials are not the dangerous storm outside, but they are the pilot's only source of the information required to navigate the storm safely, without crashing the plane. It would be crazy to disregard the dials just because they aren't the storm, wouldn't it?

"Isn't Analytic Idealism unfalsifiable?"

Before directly answering this question, it's important to notice that, when Karl Popper proposed falsifiability as a requirement for scientific theories, he was talking about, well, *scientific* theories—i.e., theories that model and predict the *behavior* of nature, not what nature *is*. A scientific theory must be falsifiable in the sense that it must make predictions about nature's future behavior that can be checked against experimental outcomes. If this is not the case, the theory is unfalsifiable and, therefore, not a proper scientific theory.

But when it comes to Analytic Idealism—and mainstream Physicalism too—we're not talking about a scientific theory that predicts nature's future behavior; instead, we're talking about metaphysical statements about what nature *is*. The criteria for choosing the best theory in this case is more diverse than falsifiability: they entail internal logical consistency, contextual coherence, conceptual parsimony, explanatory power, and empirical adequacy. The latter criterion means that the *implications* of a proper metaphysical theory must not contradict established science. And insofar as established science is falsifiable, a metaphysical theory must indeed relate to falsifiability, but only in an indirect way.

The applicable question is thus whether Analytic Idealism is *consistent with established science*. And the answer is an overwhelming 'yes.' As we have discussed earlier, established science has shown that—short of *unfalsifiable* theoretical fantasies for which there is no positive evidence—physical entities do not have standalone existence, being instead a product of measurement. This is exactly what Analytic Idealism maintains, since all 'physical' entities are dashboard representations of measurements, which only endure while a measurement is being performed. And it directly contradicts mainstream Physicalism, which presupposes precisely that physical entities, for being fundamental, *must* have standalone existence independent of observation.

Established science has also shown that there are cases—such as during the psychedelic state, as discussed earlier—in which brain activity decreases, while the richness and intensity of experience increases. This is at least very hard to make sense of under mainstream Physicalism, according to which there is nothing to experience but brain activity. But it can be comfortably accommodated by Analytic Idealism, according to which brain activity is just what inner experience *looks like*, from an external perspective; i.e., it is but an *image* of inner experience. And unlike causes, images don't need to be complete: they don't need to reveal everything there is to know about the phenomenon they represent. In the case of psychedelics, the images leave quite a bit out.

Moreover, psychedelics are only one case in which, contrary to physicalist expectations, brain function and the richness of experience are inversely correlated. As we have seen above, constriction of blood flow to the brain due to strangulation or G-forces—which reduce brain metabolism due to oxygen deprivation, or hypoxia—can lead to psychedelic-like trances and "memorable dreams." Hyperventilation—which also constricts blood flow to the brain because it induces high blood alkalinity levels—can lead to life-transforming insights, a phenomenon leveraged by some therapeutic breathwork techniques. Even outright brain *damage* can lead, in some specific cases, to richer inner experience. In a condition called 'acquired savant syndrome' (look it up), some people who have suffered brain damage because of head trauma incurred during car accidents, lightning strikes, and even bullet wounds to the head, suddenly manifest extraordinary cognitive skills such as artistic talents, the ability to perform complex calculations almost instantaneously, and perfect memory. A large group of Vietnam war veterans who suffered damage to the frontal or parietal lobes has also been shown to have a higher propensity to life-transforming religious experiences (see: "Neural correlates of mystical experience," by Irene Cristofori *et al.*, published in *Neuropsychologia*, 2016). Even patients who suffered brain damage because of surgery for the removal of tumors experience significantly higher "self-transcendence" (see: "The spiritual brain: Selective cortical lesions modulate human self-transcendence," by Cosimo Urgesi *et al.*, published in *Neuron*, 2010). Moreover, a group of so-called 'trance mediums' displayed significantly reduced activity in areas of the brain related to reasoning and language processing, precisely when engaged in activities that require high reasoning and language processing (see: "Neuroimaging during trance state: A contribution to the study of dissociation," by Julio Fernando Peres *et al.*, published in *PLoS ONE*, 2012).

I could go on and on, but you get the picture. Although most of the times brain activity directly correlates with the richness of inner experience, in some specific but broad and consistent cases the opposite is true. These cases are the black swans that disprove Physicalism and substantiate Analytic Idealism.

The scientific evidence discussed above not only addresses the question of falsifiability for Analytic Idealism and mainstream Physicalism, it also provides positive *empirical confirmation* for Analytic Idealism across very different fields of science.

"But, Bernardo, it doesn't seem fair to claim that the cases discussed above—in which inner experience and brain activity are inversely correlated—are evidence for Analytic Idealism just because the latter doesn't require the dashboard images to be complete; I would expect a more specific idealist account of what is going on there."

Fair enough. If the living body is what an alter of the natural field of subjectivity looks like, then something in the body is what the *dissociative boundary itself, plus the cognitive mechanisms that enforce it,* look like; after all, the boundary is also part of the alter.

Obviously, the outer surfaces of the body—the skin, retinas, eardrums, nose mucosa, and the inner lining of the digestive tract—are part of what the boundary and its enforcing mechanism look like. But since a dissociative boundary is a *cognitive* configuration, and since humans naturally have further levels of internal dissociation—e.g., when you forget things you know, experience cognitive dissonance, or 'park' your problems before going to work—it is reasonable to expect that *some* patterns of brain activity also represent dissociative *boundaries and related enforcement mechanisms*, not just the contents of the dissociation.

The hypothesis is thus that, when psychedelics, hypoxia, or brain damage reduce brain activity or impair brain function in some way, sometimes what they are actually reducing or impairing is *a dissociative boundary and related enforcement mechanisms*. When the boundary is thus impaired, it becomes more porous, permeable, allowing experiential states that aren't represented as brain activity to be accessed by the re-representing ego. This accounts for the extra experiential richness reported.

"Bernardo, I am an energy healer and I know that you are right! Since the brain doesn't generate the mind, neuroscience is pointless, psychoactive drugs are pointless. Let us all just hold hands and sing the Kumbaya! Only psychic healing and crystals can save us!"

Well, not so fast. Under Analytic Idealism, the 'physical' brain—and its equally 'physical' patterns of metabolic activity—are what a subject's mental states look like to external observation. As such, *they convey precious information about the subject's mind*. By studying the brain—the only form of objective access we have to someone else's inner life—neuroscience deepens our understanding of the mind in a manner that complements introspection and the corresponding subjective reports. And I believe this to be critical, due to the mind's well-known tendency to deceive itself (indeed, I have often opined that the mind's 'prime directive' is to deceive itself). You may also have noticed, in this and other works of mine, that I regularly cite and quote neuroscientific papers. The reason I do so is that I, as an analytic idealist, take neuroscience's insights seriously. Introspection alone—albeit admittedly the royal path to knowing the mind—isn't reliable enough, and doesn't reach far enough into mental processes that aren't accessible from the ego.

Psychoactive drugs—just like any other 'material' thing—are just what transpersonal mental states look like when represented on a dashboard. *But these states, when absorbed by an alter, have a causal, curative, mental-to-mental effect on the alter's inner states.* Drugs don't stop being causally effective just because they are mental in essence; on the contrary: their mental essence renders it intuitively clear why psychoactive drugs have mental effects. The neurosurgeon's scalpel, too, is what a set of causally effective mental states looks like. Pills and surgery are thus still powerful medical tools. When you see a surgeon slicing into someone's brain, that image is a dashboard representation of an external mental process piercing through a dissociative boundary to influence—in a curative manner—the patient's inner mental states. Any metaphysics that contradicts the effectiveness of such a process is just wrong, since the latter is an established empirical fact.

If ever medically required during the course of my life, I will undergo neurosurgery, and I recommend that all other idealists out there do the same. Moreover, *I take psychoactive drugs myself* every day, already for years, and can personally attest to their effectiveness (if you must know, I take low doses of Amitriptyline to alleviate severe tinnitus, but the drug has had many other positive effects in my mental inner life). I don't, for a minute, pooh-pooh medical science. Neither do I see any reason, under Analytic Idealism, to disregard neuroscience. On the contrary: under Analytic Idealism, the metabolizing brain is a map to the far corners of the mind, unreachable through introspection.

"Well, Bernardo, if all science is still valid under Analytic Idealism, and there is still a world out there independent of us, then Analytic Idealism is basically Physicalism under a different label; it all boils down to the same thing."

This astonishingly shortsighted perspective is surprisingly common. If you identified with it, don't blame yourself too harshly. The reason why the perspective is shortsighted is that it wholly ignores the colossal differences in the *implications* of Analytic Idealism when compared to mainstream Physicalism. But our culture rewards quick judgment calls and, therefore, discourages the depth of thinking required to explore the implications of new ideas.

Under Analytic Idealism, your life, your metabolism, is *not* the cause or generator of your consciousness, but merely what your private mentation *looks like* from the outside; i.e., from across your dissociative boundary. Life is what the dissociation looks like. Therefore, the end of life is the end *of the dissociation*, not the end of consciousness.

The end of a dissociative process is also not the end of the mental states held within the dissociative boundary; it is merely the end *of the dissociative boundary*. This means that the mental states previously held by the alter—your lifetime of memories and insights—are released into the broader cognitive context of nature-at-large upon death. Our hard-earned memories and insights—typically the result of much suffering—are not lost upon death but, instead, become available to nature-at-large. Contrast this with the physicalist view: when you die, all your memories and insights are just lost forever, and all that suffering was for nothing. Clearly, these two scenarios aren't even remotely similar, and their differences are of great relevance to our values, to how we make our life choices, and generally experience our lives.

In addition, although Analytic Idealism preserves—arguably even strengthens—the rationale for drugs and surgery in medicine, it opens an *additional* avenue for the treatment of organic ailments: talk therapy and related practices. For under Analytic Idealism, the body is not a mere mechanism distinct from mind, but the extrinsic appearance of mental processes. Therefore, any organic ailment is, at root, a mental ailment. This doesn't mean that you can cure cancer with positive thinking—as we've discussed before, the ego complex is naturally dissociated from autonomous functions, and thus has limited causal influence on them. But it does mean that it's sensible to research whether we can reach further down into our physiology through psychological means, so to address some 'physical' ailments. This, in fact, could be the missing account of the so-called placebo effect, which under Physicalism is just a vexing anomaly. Can we *deliberately* induce the effect through psychological methods, now that a coherent metaphysical framework validates and accounts for it?

I have already explored the implications of Analytic Idealism at length in previous writings, so won't repeat all that here. It suffices to mention—as I did above—what I believe to be two of the more important ones. The invitation to you— especially if you feel tempted to regard Analytic Idealism as equivalent to Physicalism in any important sense—is to think about the different *implications* of these very different views. What changes for you if you understand yourself to be not a physical mechanism, but a *mental being*, whose mental contents and core subjectivity will never be lost to nature?

"Yeah, okay, Bernardo. But your Analytic Idealism, contrary to my expectations, still doesn't validate my belief that I can change reality through my morning affirmations, that the universe is directly responsive to my wishes and whims, and that I can heal myself of any ailment just through positive thinking; and I really want to believe all that!"

Well, sorry then. But our quest here is an honest pursuit of truth—or at least of the best account of it that we can humanly muster. This book is not aimed at validating gullibility. And if you *truly* understand the system we are outlining here, eventually you will realize that the ego's power fantasies are *utterly unimportant*. Your life isn't, has never been, and will never be, about you—i.e., your ego. Instead, it's about *nature*. You are just a dissociated alter of nature, something nature is temporarily doing. If you really grok this, in the core of your being, your ego's need for control will melt like butter under the sun. You will understand that the ego is just a tool of nature. It doesn't need to have control of anything; it doesn't even need—uncomfortable as it may sound—to have free will. Its job is to *observe, take notice, realize, understand, metacognize, re-represent* the dance of nature; and to do so from a seemingly external perspective that, without dissociation, isn't available to nature. We're spies for the mind of nature. See this, and all your existential angst will pop like a soap bubble, hardly leaving any trace behind. True freedom and fulfillment lie in understanding what's going on, not in controlling it.

Every spring I contemplate the blossoms of my apple tree. Sometimes I fantasize about one of the blossoms thinking that its life is about *it*. Because it doesn't understand what's going on, it feels that it needs to control the whole tree and live forever as a blossom. If it would have it its way, there would be no more apples or apple trees. But if, instead, it understood its true role and importance, it would relax in that understanding and let go of control fantasies.

Unlike Physicalism—under which consciousness is but a causally-ineffective by-product of matter, an inconsequential and ephemeral anomaly—Analytic Idealism renders our conscious inner lives not epiphenomena, but *true segments of nature, alike the rest of nature in their very mental essence*. 'Behind' or 'underlying' the rest of the 'physical' universe there is something as experiential as us, whose core subjectivity we share. We belong here. We are at home. We aren't inconsequential anomalies. We aren't essentially different from the world around us. We have a role to play; one only we can play. *This* is what is important, not the validity of our fantasies of control.

Control is only a thing for those drowning in the utterly unjustified and unnecessary sea of nihilism implied by Physicalism. For those, the need for egoic control is a desperate move meant to compensate for the absence of meaning. Clinical psychology technically calls it 'fluid compensation.' But you don't need to swim in that sea anymore. So relax and enjoy the ride. Your core subjectivity is the subjectivity of nature-at-large, so what you *really* are isn't going anywhere; ever. Where *could* your core subjectivity possibly go, since it is all there is? Where is it going to disappear into? What you truly are isn't in danger; it has never been and will never be. So what is all this control silliness about?

"Let us go back to the business of death. You said that the body is what a dissociative alter in the mind of nature-at-large looks like from the outside. You also said that death is merely the end of the dissociation. But when we die, the body doesn't just disappear; it stays behind for quite a while! How can the image of the dissociation stay when the dissociation itself ends?"

The first thing to notice here is that, at the very moment the death process is completed, something very significant changes in the body: it no longer metabolizes. Indeed, that's the very definition of death. The appearance or representation of an alter of nature-at-large is a *metabolizing body*, not just the body's anatomy; it's an active, dynamic image, not just a static outline. The end of metabolism thus betrays the end of the process of dissociation, as far as dashboard indications go.

Nonetheless, a non-metabolizing body does stay behind, sometimes for quite a while. What does *that* mean, under

Analytic Idealism? Well, notice that every process in nature leaves vestiges behind after it ends; vestiges that can linger. After the rain stops the ground remains wet for a while. The doings of nature leave 'footprints' behind. This can be understood almost literally: to step on a patch of mud is a doing; but even after you stop doing it—by lifting your foot off the patch of mud—a vestige of your previous doing lingers. The dead body is entirely analogous to this: after nature stops doing a dissociative alter, the imprint of what it was doing before lingers. That 'footprint' left behind is the dead body.

"The question is, what is the *'me'* that I identify with? I identify with the individual that lives and dies, who goes by my name, not with this impersonal over-mind of nature you've been talking about. If I understand you correctly, my individuality, my personhood, ends with death, since that individuality is a result of the dissociation. Is this correct?"

Yes, it is correct. Bernardo Kastrup, as an individual agent, ends with his death. This is pretty much a matter of course, since the metabolizing body is tightly associated with my individuality under both Physicalism and Analytic Idealism, and death *is* the end of the metabolizing body. To expect our individual agency to survive death is rather precarious. When we take psychedelics and our brain metabolism is merely *reduced*, we already lose our sense of individual self in a well-known experience called 'ego dissolution.' Now imagine what happens when *all metabolism stops!* To expect that you can merrily go on as an individual at that point is quite a stretch. Something major must change in our state of consciousness upon death; something commensurable with the major 'physical' change indicated on the dashboard as the cessation of metabolism.

But let us now think this through, to see if this conclusion really implies the end of what we really are, from a first-person perspective.

When a DID patient is cured, all alters are reintegrated into the host personality. At that point, the host remembers the memories of the alters, and realizes that each and every alter was the host all along. *The patient doesn't mourn the death of the alters,* even though the alters indeed come to an end at the moment of reintegration. The reason why the host doesn't mourn the death of the alters is that the illusion of dissociation is seen through, and the reality of each alter's identity becomes evident: the dead alters weren't actual entities, but mere *doings* of the host. Think of them as a fist: when you open your hand, the fist is gone, yet nothing has been lost. For the fist was a *doing,* not an entity. The end of the dissociation is like the opening of the fist: something does change, but nothing is lost. For precisely this reason, nothing is lost when the alter that goes by your name comes to an end at death.

Here's another example. When you dream while asleep, you experience a form of dissociation: you identify only with your dream avatar, but not with the rest of the dream. You don't think that you are the buildings, trees, and people you see around you during the dream. Nonetheless, the buildings, trees, and people you see are being done by you, the dreamer—who else? When you wake up, the dissociation comes to an end and your dream avatar is toast. Yet, you don't start crying and mourning the death of your dream avatar upon awakening. Why not? Because the end of the dissociation is structurally paired with a *seeing through* the illusion: immediately upon waking up, you know that your dream avatar wasn't an entity, but just something *you* were doing; it was *you* all along, and you didn't stop existing because you stopped doing your dream avatar. It is thus reasonable to infer that, when the dissociative alter that goes by your name comes to its end at death, and the true you wakes up, you won't mourn the death of your waking-life avatar. And you will still have its memories and insights, along with its core subjectivity.

"If my true self survives death, does it mean that I will become conscious in the Otherworld?"

As far as reason and evidence suggest, *this* world is the only world; pending some non-anecdotal evidence, everything else is a fantasy. So when you die you don't go to any other world; you just leave the cockpit and become the sky outside the airplane (the sky where the airplane already was all along!). Both the cockpit and the sky outside are parts of *this* one world. The 'Otherworld' is *this* world; just experienced from a first-person—as opposed to a second- or third-person—perspective. In summary, when you die you go nowhere; you stay right in the 'place' where you are right now, although you will experience it very differently.

Death is but a change of *perspective*: the transition from *perceiving* the world to *being* the world; from *representing* the world to knowing the world *as it actually is*; from experiencing the world from a second- or third-person perspective to a first-person one. The image of a dead body decomposing and being reabsorbed into the ground is a perfect metaphor for what must happen from a first-person perspective: as the dissociation that separates us from the world comes to an end, we—our entire mental inner life—are reabsorbed into the vast experiential field that we, in life, used to inhabit and represent in the form of the 'physical' world. The end of the dissociation is a reabsorption of our inner life into the surrounding cognitive field. And all this happens in *this* world right here.

"Science hypothesizes that the universe began at the moment of the Big Bang. There were no living beings at that point, and then for billions of years later. According to Analytic Idealism, with no living beings there are no dashboards. And without dashboards there is no 'physical' universe. So how could the Big Bang have happened, since it is supposed to have been a physical event?"

The mental event that would have been represented 'physically' as the Big Bang, had there been someone there to observe it, *did happen* (assuming that Big Bang theory is correct, of course). The event-in-itself—which is always mental—happens whether it is measured and 'physically' represented or not; the states of the sky exist whether there is an airplane measuring and representing them on a dashboard or not. The entire history of the universe prior to the emergence of life did happen; it just wasn't represented 'physically,' for there was no life to represent it. Had there been someone there to look at it, then the history of the universe prior to life would have looked like what our physical models predict it did.

The same rationale applies to whatever aspects of the 'physical' universe aren't observed right now. For instance, there is no life visually observing the core of planet earth as you read these words. Therefore, the 'core of planet earth,' as a 'physical' thing, does not exist right now; representations don't exist unless something is being represented. But the thing that would be represented as the 'core of planet earth' if someone were there looking at it right now, that *mental* thing does exist right now. It just isn't 'physical,' since it is not being represented.

"Earlier you talked about the similarities between the network topology of mammalian brains and the cosmic web of galaxies. You used these similarities to substantiate your claim that, just like the brain is the extrinsic appearance of our conscious inner life, so is the universe-at-large the extrinsic appearance of some universal consciousness. But it took hundreds of millions of years, after the Big Bang, for there to be a web of galaxies. Before that, the universe didn't look like a brain at all. So how could it have been the outer appearance of a mind then?"

At the moment of conception, when a human being—or any mammal, for that matter—begins its life in the form of a single-celled zygote, the human being also has no brain. Just like the universe after the Big Bang, it also takes a human being quite a while—many weeks—to develop the network topology we call a brain. Does that mean that a fetus has no mind? Of course not. It only means that it doesn't yet have as complex a mind as a fully formed human does. The evolving complexity of the human nervous system during gestation 'reflects'—i.e., is the dashboard representation of—the evolving inner differentiation or complexification of our mind; i.e., of the contents and inner cognitive topology of our alter. In just the same way, it stands to reason that the slow evolution of the cosmic web of galaxies, for hundreds of millions of years after the Big Bang, reflects the slow inner differentiation or complexification of the mind of nature-at-large.

"Does this mean that the mind of nature is highly evolved and complex, featuring high-level mental functions such as self-awareness, metacognition, etc.?"

No. Evolution here should be understood as a relative term: the formation of the cosmic web of galaxies suggests that the mind of nature *today* is more complex than it was *just after the Big Bang*. But this doesn't mean that it is as complex today as the dissociated mind of an alter—such as you and me—is today. We must compare the present evolutionary state of a mind with its own earlier state, not with the evolutionary state of its dissociated centers of awareness. We don't have suitable common references for the cross-comparison, as the behavior of the universe as a whole is incommensurable with the behavior of mammals walking around planet earth.

We evolved in the context of a planetary ecosystem with finite resources. The concomitant evolutionary pressures forced us to develop complex, reactive minds, which could respond to changing environmental challenges and opportunities. It is for this reason that we have high-level mental functions such as self-awareness and metacognition. All evidence indicates overwhelmingly that less evolved organisms do not have these high-level functions. Although the behavior of many single-celled organisms is complex enough to betray the presence of consciousness—amoeba, for instance, build little shells out of mud particles to protect themselves—it certainly doesn't suggest the presence of any high-level mental function. We don't see insects staring into a mirror, dazed by self-recognition, or cats walking around pondering the meaning of their lives. High-level mental functions are thus *not* inherent to the field of subjectivity that nature is—otherwise every living being, even bacteria, would have them—but evolved within the constraints of a planetary ecosystem.

The mind of nature-at-large did not evolve within a planetary ecosystem; it was never subjected to the evolutionary pressures that cause the development of high-level mental functions. Hence it doesn't have such functions. Even the predictability of the laws of nature suggests a simple, instinctive, spontaneous mind, not a reactive one engaging in deep introspection, premeditation, planning, pondering, etc.

"Why did the mind of nature decide to dissociate itself to begin with?"

It almost certainly didn't decide anything, as an explicit decision requires high-level mental functions such as metacognition and the ability to deliberate. As such, there may have been no reason—no explicit motivation—for the first dissociation. For everything that *can* happen in nature, given enough time, eventually *does* happen. And since dissociation obviously can happen—otherwise we wouldn't be having this conversation—it should come as no surprise that it eventually did.

Notice that the problem of how or why the mind of nature became dissociated to begin with *is* the problem of abiogenesis, the origin of life from non-life. After all, life is merely what the dissociation looks like. So to ask why the mind of nature dissociated is the exact same question as to ask why life arose: *it probably just happened.* And in the case of the first dissociation—i.e., abiogenesis—once it happened the forces of natural selection made sure that the resulting alters developed means to maintain the dissociation—i.e., to survive—and reproduce, thereby spinning off more alters. The alters survived because, at some point, an alter arose that *could* survive; and the alters generate more alters because, at some point, an alter arose that *could* reproduce.

"But in humans, DID almost invariably happens as a response to trauma. So isn't it conceivable that the mind of nature, in its original non-dissociated state, suffered some kind of trauma?"

It is indeed conceivable, but the question is whether entertaining this line of speculation is productive. With humans, trauma is triggered by outside agents; it isn't an endogenous condition that arises purely from inside. And since there was nothing outside the mind of nature in its original, non-dissociated state, there couldn't have been external triggers.

Can we imagine that dissociation could still have been triggered *internally*, by some kind of traumatic 'thought'? Yes. We can speculate that, upon intuiting its indelibly lonely condition at some point in its process of inner complexification, the mind of nature would have had plenty of emotional impetus to spontaneously dissociate. But this line of speculation projects too much of human psychology onto something utterly nonhuman; it anthropomorphizes nature too much. And even if it didn't, it would still be an arguably unproductive line of inquiry, in that we can't base it on any kind of objective evidence.

Ultimately, nature dissociated because dissociation was possible; and that's all we can say with confidence.

"This dissociation business sounds like you are pathologizing nature, portraying it as sick, or crazy, since DID is a pathology."

The claim is not that the mind of nature has a disease called DID; the claim is that it undergoes a dissociative process in some sense *akin* to DID. We, humans, nominally define DID as a pathology because it is usually dysfunctional in the context of human activity. But the criteria for this, of course, don't apply to nature-at-large. Arguably, nature is even *more functional* with something akin to DID, if we take the variety of life to be something that enriches nature by creating complexity despite the second law of thermodynamics (although this, too, is anthropomorphizing things). Dissociation is just a natural process; something that spontaneously happens in nature, like black holes, quasars, supernovae, etc. To think of it as a disease just because humans consider it dysfunctional in human society is not a valid logical step.

"Can we call the mind of nature God?"

Well, it depends on what we mean by 'God.' If God is supposed to be, or have, a deliberate mind external to nature, which interferes with nature's business from the outside, then surely not. Under Analytic Idealism the mind in question *is* nature; there is no external mind interfering with nature from the outside, for there is nothing outside nature.

Nonetheless, I acknowledge that, under Analytic Idealism, the mind of nature—which is what nature *is*—is omnipresent, omniscient, and omnipotent. It is omnipresent because, since nature *is* this mind, the mind is necessarily everywhere in nature. It is omniscient because the defining characteristic of a mind—of a field of subjectivity—is that it is sentient. And it is omnipotent in the sense that, whatever happens in nature, happens necessarily because of an action—an excitation, a 'movement,' an 'oscillation,' a 'ripple'—of the field of subjectivity that nature is. If these three attributes are sufficient, in your view, to qualify the field of subjectivity as divine, then I suppose it's okay to call it 'God.' I have done it myself a few times, though mostly metaphorically. Just remember that other capacities commonly attributed to God in Western culture—self-awareness, wisdom, the ability to act deliberately, to have a plan, etc.—are not entailed or implied by Analytic Idealism; on the contrary.

"What is this all for? I mean, life, the universe, and everything? What's the meaning of life?"

To presume that nature needs a deliberate reason to do what it does is, again, to anthropomorphize nature. Nature does what it does because it is what it is; its behavior is an implication of its being and intrinsic properties. In simpler words, because nature is what it is, it cannot help but do what it does. Its actions are spontaneous, 'instinctive,' not deliberate—at least this is what all evidence seems to point to.

Which is not to say that life is meaningless. Whether nature has a plan or not, it is coherent to imagine that we may find ourselves having a meaningful, productive role in it. As a matter of fact, I think we can go even further and identify what that role is.

In the bloody course of the evolution of life on earth, the more advanced organisms have developed high-level mental functions. Humans, in particular, seem to be unique in our ability to think symbolically, conceptually. Other higher animals—such as cetaceans, pachyderms, apes, and perhaps even some mollusks—also seem to have some degree of self-awareness. If the four-billion-year-long evolutionary drama is pushing towards something, it seems to be these high-level mental functions.

Now notice that it is *only* through these high-level functions that nature can take explicit notice of itself; raise its head above the tsunami of instinctual unfolding and take account of what it is doing; perhaps even of what it is. It is only through life—through dissociation—that nature can 'step out of itself,' so to contemplate itself with some degree of objectivity. As Jung put it, this meta-cognitive scrutiny is a second act of Creation, for it bathes existence with the light of a new level of awareness.

There is a sense, thus, in which we are 'spies for God.' We are in the unique position, after the unfathomable labor of four billion years of evolution, to contemplate nature from a vantage point not otherwise available to nature. Countless conscious beings have lived and died over countless eons, so we could stand here today, musing about the most profound questions of existence. And after a lifetime of insights in this regard, upon death—the end of the dissociation—we contribute those insights to the broader field of cognition that nature is.

It is difficult to imagine that this *isn't* a meaningful role, whether the arrangement was deliberate or not. The ancients seem to have intuited this, since they chose to symbolize death as an agent—the grim reaper—wielding a *harvesting* instrument, of all things. Even more remarkably, they also considered sacrificial offerings gifts to God. Why would they think that? Why would the end of a life give God something It wanted or needed? Shortsighted and morally unacceptable as sacrifices are—we're all going to die anyway, so speeding up the process just wastes learning opportunity—the idea does seem to reflect a deep, spontaneous intuition about the value of life, and of death as the means for nature to bank or collect on that value.

Chapter 8

Time, space, identity, and structure

As we've seen, according to Analytic Idealism all there exists in nature is *one* subject—one field of subjectivity—whose excitations are, well, *everything experienced*. The whole of nature *is* that one subject.

But how can one universal subject be you, me, everybody else, and everything else at once? This is perhaps the most difficult aspect of Analytic Idealism to wrap one's head around, for it implies that you are me, at the same time that you are yourself. How can this possibly be? After all, you can see the world through your eyes right now, but not through mine.

Although reference to dissociation, empirically validated as it is, forces us to accept that this somehow can indeed be the case—for it *is* the case in severely dissociated human minds—the question of how to *visualize* the mechanisms of dissociation remains difficult. How can you visualize a process by virtue of which you are me while being yourself concurrently? How are we to get an intuitive handle on this?

Notice that what makes it so difficult is the *simultaneity of being* implied by the process. You can easily visualize yourself being your five-year-old self—an entity different from your present self in just about every way—because being your five-year-old self is not *concurrent* with being your present self: one is in the past, the other is in the present. Visualizing oneself taking two different points of view into the world does not offer any challenge to our intuition, provided that these points of view aren't taken concurrently.

Here is another example. When I was a child, I used to observe a very curious behavior of my father's: he liked to play chess against himself, a common and effective training technique at a time before computerized chess engines. Doing so helps a chess player learn how to contemplate the position on the board from the opponent's point of view, so to anticipate the opponent's moves. My father would perform this exercise quite literally: he would play a move with the white pieces, turn the entire board around by 180 degrees, then play a move with the black pieces, after which the process would repeat itself.

My father—a single subject—was taking two different points of view into the world, experiencing the battle drama of the game from each of the two opposing perspectives; one subject, two points of view. We have no difficulty understanding this because the two perspectives weren't simultaneous, but instead occupied *distinct points in time*.

Yet, we've known for over a century now—since the advent of Einstein's theory of General Relativity—that time and space are merely facets of one and the same thing: the fabric of *spacetime*. Both are dimensions of extension in nature, which allow for things and events to be distinct from one another by virtue of occupying different points in that extended fabric. For if two ostensibly distinct things occupy the same point in both space and time, then they can't actually be distinct. But a difference in location in either space or time suffices to create distinction and, thereby, structure. By occupying the same point in space, but at different times, two objects or events can be distinguished from each other. If they exist simultaneously but at different points in space, they can also be distinguished.

The way to gain intuition about how one subject can seem to be many is to understand that *differences in spatial location are, in an important sense, physically equivalent to differences in temporal location.* This way, for the same reason that we have no difficulty in intuitively understanding how my father—a single subject—could seem to be two distinct chess players, we should also have no intuitive difficulty in understanding how one universal subject can be you and me: just as my father could do so by occupying different perspectives at different points in *time*—that is, by alternating between black and white perspectives—the universal subject can do so by occupying different perspectives at different points in *space*; for, again, space is essentially the same thing as time. That you and I always occupy different points in space is the reason why you can be me, while still being you.

Yet, the demand for this transposition from time to space still seems to be too abstract, not concrete or intuitively satisfying enough; at least to me. We need to make our analogy a little more sophisticated.

A few years ago, I had to undergo a simple, short, but very painful medical procedure. So instead of pumping me full of opioids, the doctors decided to give me a fairly small dose of a general anesthetic, which would knock me out for about 15 minutes. I figured that this would be a fantastic opportunity for an experiment: I would try to focus my attention introspectively and fight the effects of the drug for as long as I could, so to observe the subjective effects of the anesthetic. I had undergone general anesthesia before, in my childhood, but had no recollection of it, so this was a fantastic chance to study my own consciousness with the maturity and deliberateness of an adult.

And so there I was, lying on an operating table, rather excited about my little experiment. The drug went in, intravenously, and I turned my attention to the contents of my own consciousness. Yet, as the seconds ticked by, I couldn't notice anything. "Strange," I thought, "nothing seems to be happening." After several seconds I decided to ask the doctors if it was normal for the drug to take so long to start causing an effect. Their answer: "We're basically done, just hang in there for a few more moments so we can wrap it up."

"What?!" I thought. "They are basically done? How can that be? It hasn't been a minute yet!" In fact, more than 15 minutes had already elapsed; they had already performed the whole procedure. I experienced absolutely no gap or interruption in my stream of consciousness; none whatsoever. Yet, obviously there had been one. How could that be? What had happened to my consciousness during the procedure?

The drug had altered my perception of time in a very specific and surprising way. If we visualize subjective time as a string from where particular experiences—or, rather, the memories thereof—hang in sequence, the drug had not only distorted or eliminated access to some of those memories, but also cut off a segment of the string and tied the two resulting ends together, so to produce the impression that the string was still continuous and uninterrupted. I shall call this peculiar dissociative phenomenon 'cognitive cut and tie': access to the memories of certain experiences in a cognitively associated line is removed from the line, and the two resulting ends seamlessly re-associated together, so the subject notices nothing missing.

Now let us bring this to bear on my father's chess game. Imagine that we could manipulate my father's perception of time in the following way: we would cut every segment of time when my father was playing white and tie—that is, cognitively associate—those segments together in a string, in the proper order. We would also do the same for the black segments. As a result, my father would have a coherent, continuous memory of having played a game of chess only as white, and another memory of having played another—albeit bizarrely identical—game of chess only as black. In both cases, he would presume his opponent to have been somebody else. If you were to tell my father that it was him, himself, on the other side of the board all along, he would have thought you mad. For how could the other player be him, at the same time that he was himself?

The answer to how one universal subject can be many—to how you can be me, as you read these words—resides in a more sophisticated understanding of the nature of time and space, including the realization that, cognitively speaking, what applies to one ultimately applies to the other. As such, if you believe that you were your five-year-old self, then there is an important sense in which, by the same token, you must believe that you can be me right now. There is only the universal subject, and it is *you*. When you talk to another person, that other person is just you in a 'parallel timeline'—which we refer to as a different point in space—talking back to you across timelines. The problem is simply that 'both of you' have forgotten that each is the other, due to 'cognitive cut and tie.' Do you see what I mean?

The hypothesis here is that, subjectively speaking, *a different position in space is just a different point in a multidimensional form of time, and vice versa.* Indeed, such interchangeability between space and time is a field ripe with speculation in physics. Physicist Lee Smolin, for instance, has proposed that space can be reduced to time. Physicist Julian Barbour, in turn, has proposed the opposite: that there is no time, just space. There may be a coherent theoretical sense in which both are right.

The most promising theoretical investigation in this area is perhaps that of Prof. Bernard Carr, from Queen Mary University of London. If his project is given a chance to be pursued to its final conclusions, it is possible that physics will offer us a conceptually coherent, mathematically formalized way to visualize how one consciousness can seem to be many. Looking upon personal identity through the lens suggested above may convince you that, when an old wise man turns to a brash young lad and says, "I am you tomorrow," such statement may have more layers of meaning than meets the eye at first.

This is not to say that, if my free speculations above turn out to be false—or even merely implausible—Analytic Idealism is refuted or wounded; it isn't. That dissociation can make one subject seem to be many, concurrently conscious other subjects is an *empirical fact*; it does happen in nature and we know it. Whether we have a conceptually clear, explicit, and satisfying account of this fact or not, the fact remains. And the fact is all that is required to substantiate Analytic Idealism. My goal with these brief speculations is merely to give you *one* way to imagine how dissociation might operate—so it doesn't look so conceptually unapproachable to you anymore—not to make the speculations an integral element of the argument for Analytic Idealism.

My little experiment on the operating table and its conclusion—the 'cognitive cut and tie' mechanism—show how inherently *subjective* time and space are: a rather trivial interference with my inner mental states seamlessly eliminated a whole segment of time, without leaving any noticeable vestige or loose ends behind. And sure enough, this is an implication of Analytic Idealism: since the 'physical' world is an alter's internal representation of the *real* world out there, so must time and space be; for time and space are the dimensions of the 'physical' world—of the screen of perception—which is internal to the alter.

Analytic Idealism is not alone here. Since at least the late 18th century, a series of developments in Western philosophy and science—such as Kant's and Schopenhauer's assertion that spacetime is a category of perception, Einstein and his Block Universe, Julian Barbour and his universe without time, Lee Smolin and his universe without space, Loop Quantum Gravity, the cognitive psychology of temporal perception, etc.—have relegated spacetime to the status of a "stubbornly persistent illusion," in the words attributed to Einstein, as opposed to the fundamental scaffolding of reality.

The problem is that, if spacetime exists only in the alters and not in the *real*, external world, then the external world has no extension. And extension seems to be a prerequisite for differentiation and structure. After all, it seems inevitable to conclude that things and events can only be distinguished from one another insofar as they occupy different volumes of space or different moments in time. Without spatiotemporal extension, all of nature would seem to collapse into a singularity without internal differentiation and, therefore, without structure. Schopenhauer had already seen this in the early 19th century, when he argued that spacetime is nature's *"principium individuationis,"* or 'principle of individuation.'

Nonetheless, it is empirically obvious that nature *does* have structure: its various regularities of behavior betray just that. Under certain circumstances nature does one thing and, under others, something else; repeatedly and reliably. Such distinguishable but consistent behaviors can only occur with some form of underlying, immanent structure.

So how are we to reconcile the empirical fact that nature has structure with the emerging Western understanding that spacetime is not fundamental? How are we to think of the irreducible foundations of nature as both lacking extension *and* having structure? I submit that this is the least recognized and discussed dilemma of modern science; and one that Analytic Idealism resolves.

To understand the solution, we must start with an admission: 'physical' objects and events *do indeed* inherently require spatiotemporal extension to be differentiated; Schopenhauer was right about the *principium individuationis*. But we know of one other type of natural entity whose intrinsic structure does not require extension.

Consider, for instance, a hypothetical database of student records. Each record contains the respective student's intellectual aptitudes and dispositions, so the school can develop an effective educational workplan. The records are linked to one another so to facilitate the formation of classes: students with similar or compatible aptitudes and dispositions are associated together in the database. Starting from a given aptitude, a teacher can thus browse the database for compatible students.

Now, notice that these associations between records are fundamentally *semantic*: they represent links of *meaning*. Associated records *mean* similar or compatible aptitudes, which in turn *mean* something about how students naturally cluster together. Therein lies the usefulness of the database. Even though the latter may have a spatiotemporal *embodiment*—say, paper files sorted by box in an archive—there is a sense in which their structure fundamentally resides in their *meaning*. Spatiotemporal embodiments merely *copy* or *reflect* such meaning. After all, the semantic relationships between my intellectual aptitudes and those of others wouldn't disappear just because our respective paper files went up in flames.

I submit that this is how we must think of the most foundational level of nature, the universe behind extension: as *a database of natural semantic associations*, spontaneous links of meaning, *cognitive associations*. This is analogous to how a mathematical equation associates variables based on their meaning, whether such associations happen to have spatiotemporal embodiments or not. The associations can indeed be *projected* onto spacetime—just as databases can have 'physical' embodiments—but, in and of themselves, they do not require spacetime to be said to exist. This is how nature can have structure without extension.

But what about causality? Its central tenet is that effect follows cause in time, so what are we to make of it without extension? Philosopher Alan Watts once proposed a metaphor to illustrate the answer: imagine that you are looking through a vertical slit on a wooden fence. On the other side of the fence, a cat walks by. From your perspective, you first see the cat's head and then, a moment later, the cat's tail. This repeats itself consistently every time the cat walks by. If you didn't know what is actually going on—i.e., the existence of the complete pattern called a 'cat'—you would understandably think that *the head causes the tail.*

Behind extension, the universe is the complete pattern of semantic or cognitive associations—i.e., the complete cat. Our ordinary traversing of spacetime is our looking through the slit in the fence, experiencing partial segments of that pattern. All we see is that the cat's tail consistently follows the cat's head every time we look. And we call it 'causality.'

The notion that, at its most fundamental level, nature is a complete pattern of cognitive associations has been hinted at by physicists before. Max Tegmark, for instance, has proposed that matter is mere "baggage," the universe consisting purely of abstract mathematical relationships (see his book, *Our Mathematical Universe*, 2015).

We must, however, avoid vague, abstract hand-waving: every mathematical structure ever devised has existed in a *mind*, not in a vacuum. Insofar as they are to have meaning, the only coherent and explicit conception of mathematical objects is that of *mental* objects. To speak of mathematical structure—or of meaning—without a mind is like speaking of the Cheshire Cat's grin without the cat. Unless you are Lewis Carroll, you won't get away with it.

Meaning—such as those of the variables in a mathematical equation—is an intrinsically *mental, cognitive* phenomenon. In the absence of spacetime, this betrays the only possible metaphysical ground for a universal semantic database: *the universe is a web of semantic associations in a field of spontaneous, natural mentation;* for mind is the only metaphysical substrate we know of that isn't extended.

Indeed, mental dispositions and aptitudes are palpably real—in the sense of being known through direct acquaintance— yet transcend extension. What is the size of my aptitude for math? What is the mass or electric charge of my disposition to philosophize, or even of my next thought? Whatever theory of mind you subscribe to, the pre-theorical fact remains: you can't take a tape measure to my next thought; mentation is not extended.

As such, within the bounds of coherent and explicit reasoning, a structured universe without irreducible extension is necessarily a *mental* universe—not in the sense of residing in our individual minds, but of consisting of a field of natural, spontaneous mental activity, whose intrinsic 'dispositions' and 'aptitudes' present themselves to us as the 'laws of nature.'

This way, by coherently articulating how reality can be entirely mental—i.e., experiential—Analytic Idealism provides sound metaphysical ground for the emerging understanding, in both science and philosophy, that spacetime is not fundamental. Analytic Idealism accounts for how the universe can have inherent *structure* without extension, 'law' without dimension. Without it, we would have to follow Lewis Carroll in stating, with a straight face, that the Cheshire Cat's grin can stay behind after the cat disappears.

In conclusion, the emerging scientific understanding that space and time aren't the immutable, fundamental scaffolding of reality—that, instead, space and time are malleable, interchangeable, epiphenomenal—accounts for how dissociation creates the illusion of multiple individual identities, even though there is only one true subject in nature. This gives explicit substantiation to one of the more challenging ideas behind Analytic Idealism. And as if to reciprocate, Analytic Idealism, in turn, provides coherent metaphysical ground for the emerging scientific understanding that spacetime isn't fundamental: a *mental* universe accounts for how the universe can still have structure and 'law' in the absence of irreducible extension.

Chapter 9

Wrap-up and outlook

Much more could be said about Analytic Idealism, the philosophical argument that underpins it, the empirical evidence for it, its symbiotic relationship with modern science, and its immense implications for our lives. But this book is meant as a compact—albeit thorough—summary; something you could read in a weekend. Its brevity is not an arbitrary or commercial choice, but integral to the effect I intended the book to have: to convey a wholistic understanding of Analytic Idealism; to help you see how the whole thing comes together, as opposed to missing the forest for the trees. Whether I succeeded in this endeavor only you can judge. For those who require a more in-depth, rigorous, or detailed treatment of the topics addressed, I recommend my previous works, be them books, essays, academic articles, or audio-visual courses and talks. Much of this is freely available online.

I want to leave you with a thought that only the next generations of scientists and philosophers will be able to fully engage with: once we are acclimatized to Analytic Idealism, we realize that *semantic associations and dissociations are the primary processes*—not entities, *processes*—*of nature*. Semantic associations give nature its structure in the absence of extension; a structure that manifests itself to our observation as the 'laws of nature,' which were responsible for the creation of complexity and differentiation in a universe that, at the moment of the Big Bang, had neither. These semantic associations, inherent to the very fabric of reality, are the 'cosmic egg' that hatches and grows into the cosmic web of galaxies.

Dissociation, in turn, accounts for life itself. After all, life *is* dissociation. Without dissociation, there is a very important sense in which nature wouldn't be able to take stock of itself. Dissociation is the enabler of a second 'act of Creation,' whereby nature 'steps out of itself'—out of its own overwhelming, automatic play of instinct—and contemplates itself through our eyes. Dissociation bathes nature with the light of meta-consciousness, granting it a different form of (meta) existence.

With these two processes, we can metaphysically account for both the mind-boggling complexity of the inanimate universe, and the mind-boggling richness of life, with only one universal subject in our reduction base. Semantic association and dissociation: the double-root of all complexity. One day, scholars and investigators may be able to trace all our knowledge back to this root. A form of universal 'psychology' will then be seen as the primary science, before physics; or perhaps physics will morph into an objective natural psychology. Graph, Set and Category theories will be regarded as the primary conceptual modeling tools. And all this will confer on humanity—if we don't kill ourselves before then—a level of understanding of self and other that, today, is unimaginable to us.

The question, of course, is: *how long will this take?*

I have been writing and talking about Idealism for some 15 years now. During this period, I have participated in many debates with other scholars and scientists, as well as interviews that amounted to debates. And one change has been very noticeable to me: unlike several years ago, it has now become very hard to find physicalists willing to explicitly defend Physicalism in a debate. Even those few who are still ostensibly willing to do so, usually start the debate by—startlingly—portraying themselves as metaphysically neutral or open-minded (even when they've spent the previous four decades arguing for Physicalism). They then proceed to repeatedly ask me questions about Analytic Idealism—effectively interviewing me—instead of defending their own views. The debates I am referring to are a matter of public record, so I encourage you to watch them with what I just said in mind; it's a telling experience.

We've arrived at a juncture in our culture's history where physicalists are, by and large, only willing to be physicalists if they are engaging soft targets, such as religious fundamentalists and new agers. But when it comes down to an actual scholarly debate, suddenly they aren't really physicalists—or illusionists, eliminativists, or any other variation of Physicalism—anymore. Little reveals more palpably the fragility of mainstream Physicalism today, despite everything it has going for it in terms of cultural dynamics, as discussed earlier in this book.

I am not one to predict revolutionary, imminent changes. But slowly and progressively as the case may be, things *are changing* in our cultural landscape, when it comes to metaphysics. Twenty years ago, anything that wasn't Physicalism was automatically and sincerely considered woo-woo, mystical mumbo jumbo, wishful thinking, etc. This is no longer the case now. Despite the unfortunate intermezzo of Panpsychism, Idealism-leaning ideas are proliferating not only in philosophy, but even—perhaps mainly—in foundations of physics (think of the physics of first-person perspective) and neuroscience (think of Integrated Information Theory, which dovetails surprisingly well with Analytic Idealism).

Of course, this is not to say that Physicalism isn't trying to survive. There are just too many vested interests to protect, too many public personas to safeguard. Mainstream ideas die hard, even when they are blatantly silly ideas. And so I will offer a prediction of what Physicalism's survival effort will look like in the short-term future: physicalists will try to save face by *liberally redefining the meaning of the word 'physical' so that, whatever reality appears to be, they will just call it 'physical.'* In other words, Physicalism will become unfalsifiable by the grace of mere word redefinition. If you think this would be too silly and transparent a maneuver to try, watch with attention.

Thankfully, however, history shows us that, although bad ideas survive for much longer than they should, *ultimately* they cannot resist the steady assault of simple reason and evidence. The writing is already on the wall: all arrows are pointing to a form of objective Idealism—such as Analytic Idealism—as the next mainstream worldview. Will this new metaphysics be *the truth*? Of course not; we're apes, and apes are not in the business of ultimate truths. But it will be *less wrong*. And when all is said and done, to be less wrong is the best we can hope for.

This book has given you a glimpse of what is coming: a less wrong understanding of the nature of reality. What for you may have been a difficult and counterintuitive intellectual journey, for your great-grandchildren will be self-evident. They, and their children, will live life according to a very different understanding of what the world and the self are, what the meaning of it all is, and will have a very different sense of purpose. They will study our times and wonder how an entire culture—a civilization—could have been taken in by a story as transparently bonkers as Physicalism, for so long. Research grants will be given to investigate and understand the sociopsychology behind such a peculiar phenomenon. And we, early 21st-century people, will be regarded with the same patronizing dismissiveness with which we arrogantly regard our pre-Enlightenment ancestors.

The fact that you've read this book makes you a pioneer of the next pivoting of Western metaphysics. And for this reason, it also invests you with the responsibility of playing a social role in this pivoting. After all, there is nothing to the abstract 'collective' of a culture but *concrete individuals*; individuals such as you and me. How exactly the upcoming cultural transition will unfold, we cannot know right now. All we can be certain of is that, however it unfolds, it will do so because of *our* actions.

IFF
BOOKS

ACADEMIC AND SPECIALIST

Iff Books publishes non-fiction. It aims to work with authors and
titles that augment our understanding of the human condition,
society and civilisation, and the world or universe in which we live.
If you have enjoyed this book, why not tell other readers by
posting a review on your preferred book site.
Recent bestsellers from Iff Books are:

Why Materialism Is Baloney
How true skeptics know there is no death and fathom answers
to life, the universe, and everything
Bernardo Kastrup
A hard-nosed, logical, and skeptic non-materialist metaphysics,
according to which the body is in mind, not mind in the body.
Paperback: 978-1-78279-362-5 ebook: 978-1-78279-361-8

The Fall
Steve Taylor
The Fall discusses human achievement versus the issues of war,
patriarchy and social inequality.
Paperback: 978-1-78535-804-3 ebook: 978-1-78535-805-0

Brief Peeks Beyond
Critical essays on metaphysics, neuroscience, free will,
skepticism and culture
Bernardo Kastrup
An incisive, original, compelling alternative to current mainstream
cultural views and assumptions.
Paperback: 978-1-78535-018-4 ebook: 978-1-78535-019-1

Framespotting
Changing how you look at things changes how
you see them
Laurence & Alison Matthews
A punchy, upbeat guide to framespotting. Spot deceptions and
hidden assumptions; swap growth for growing up. See and be free.
Paperback: 978-1-78279-689-3 ebook: 978-1-78279-822-4

Is There an Afterlife?
David Fontana
Is there an Afterlife? If so what is it like? How do Western ideas
of the afterlife compare with Eastern? David Fontana presents the
historical and contemporary evidence for survival of
physical death.
Paperback: 978-1-90381-690-5

Nothing Matters
a book about nothing
Ronald Green
Thinking about Nothing opens the world to everything by
illuminating new angles to old problems and stimulating new
ways of thinking.
Paperback: 978-1-84694-707-0 ebook: 978-1-78099-016-3

Panpsychism
The Philosophy of the Sensuous Cosmos
Peter Ells
Are free will and mind chimeras? This book, anti-materialistic but
respecting science, answers: No! Mind is foundational
to all existence.
Paperback: 978-1-84694-505-2 ebook: 978-1-78099-018-7

Punk Science

Inside the Mind of God

Manjir Samanta-Laughton

Many have experienced unexplainable phenomena; God, psychic
abilities, extraordinary healing and angelic encounters. Can
cutting-edge science actually explain phenomena
previously thought of as 'paranormal'?

Paperback: 978-1-90504-793-2

The Vagabond Spirit of Poetry

Edward Clarke

Spend time with the wisest poets of the modern age and of the
past, and let Edward Clarke remind you of the importance of
poetry in our industrialized world.

Paperback: 978-1-78279-370-0 ebook: 978-1-78279-369-4

Readers of ebooks can buy or view any of these bestsellers by
clicking on the live link in the title. Most titles are published in
paperback and as an ebook. Paperbacks are available in traditional
bookshops. Both print and ebook formats are available online.
Find more titles and sign up to our readers' newsletter at
www.collectiveinkbooks.com/non-fiction
Follow us on Facebook at
www.facebook.com/CINonFiction